Republic of Numbers

REPUBLIC
of NUMBERS

Unexpected Stories of Mathematical Americans through History

David Lindsay Roberts

Johns Hopkins University Press *Baltimore*

© 2019 Johns Hopkins University Press
All rights reserved. Published 2019
Printed in the United States of America on acid-free paper
9 8 7 6 5 4 3 2 1

Johns Hopkins University Press
2715 North Charles Street
Baltimore, Maryland 21218-4363
www.press.jhu.edu

Library of Congress Cataloging-in-Publication Data

Names: Roberts, David Lindsay, author.
Title: Republic of numbers : unexpected stories of mathematical
    Americans through history / David Lindsay Roberts.
Description: Baltimore : Johns Hopkins University Press, 2019. |
    Includes bibliographical references and index.
Identifiers: LCCN 2019007917 | ISBN 9781421433080
    (hardcover : alk. paper) | ISBN 9781421433097 (electronic) |
    ISBN 1421433087 (hardcover : alk. paper) | ISBN 1421433095
    (electronic)
Subjects: LCSH: Mathematicians—United States—Biography. |
    Mathematics—United States—History.
Classification: LCC QA28 .R63 2019 | DDC 510.92/273—dc23
LC record available at https://lccn.loc.gov/2019007917

A catalog record for this book is available from the British Library.

*Special discounts are available for bulk purchases of this book. For more
information, please contact Special Sales at 410-516-6936 or
specialsales@press.jhu.edu.*

Johns Hopkins University Press uses environmentally friendly book
materials, including recycled text paper that is composed of at least
30 percent post-consumer waste, whenever possible.

For Tomás Kalmar, choicest of chums

Contents

Republic of Numbers

Introduction

The science of mathematics, both pure and mixed, can never cease to be
interesting and important to man, as long as the relations of quantity shall
exist, as long as ships shall traverse the ocean, as long as man shall measure the surface or heights of the earth on which he lives, or calculate the
distances and examine the relations of the planets and stars; and as long
as the iron reign of war shall demand the discharge of projectiles, or the
construction of complicated defences.

Benjamin Silliman

THIS BOOK BREEZES across more than 200 years of US history, through
life stories of 23 individuals who interacted significantly with mathematics. From Nathaniel Bowditch (1773–1838) to John Nash (1928–2015),
some of these individuals are famous for their mathematical activities,
some are famous for other reasons, and some will be obscure to almost
everyone. A few, mainly in the nineteenth century, had relatively brief encounters with mathematics and then moved on to other activities. This trajectory becomes increasingly rare in recent years, reflecting the growing
professionalization of the subject. A part-time or temporary affiliation with
mathematics has become unusual.

I have tried to include persons from a variety of backgrounds, and with
a variety of relationships to mathematics: pure and applied, advanced and
elementary, popular and technical. The classroom has been, and remains,
the primary venue in which Americans have encountered mathematics; or,
to be more precise, the primary venue in which they are aware of encountering mathematics. Perhaps it is not surprising, then, that most of the

subjects of this book have at some point stood in front of a mathematics class, in a school or in a postsecondary institution. But only six spent their entire careers as college professors, and of these six, three spent considerable time teaching subjects other than mathematics.

I do not, however, claim that I have come anywhere close to summarizing all manifestations of mathematical activity in the United States, or that I am providing a representative sample of mathematical Americans. I have made no more than a gesture toward representation, limited by my background, knowledge, and interests. The reader is invited to imagine, given the modest variety in this book, how much more varied and extensive a truly representative collection would be. Some of my choices will doubtless appear idiosyncratic, for they have been based not only on what may be considered objective measures of significance but also on how much curiosity these individuals generated in me personally. Choices were further influenced by whether I happened to have access to adequate knowledge of an individual. The quest for chronological diversity was another complicating selection factor. Finally, for those individuals for whom substantial biographical treatments have been published, I needed to convince myself that my point of view was distinctive enough to be worth setting in print.

Two or three of those featured herein might make a list of the greatest American mathematical minds of all time, and some can lay claim to be the first X to do Y, but such superlatives have not been a primary criterion for inclusion. I consider every person whom I discuss to be remarkable in some way, but not necessarily admirable in all respects. Encountering mathematics does not automatically produce noble souls, or even special mental facility outside of mathematics.

I have ordered the chapters by decade, based on the date of an incident or episode recounted in the opening of each chapter. But the chapters will generally range widely before and after that initial episode, in order to cover the life story. Three chapters are devoted to pairs of linked individuals.

I hope that readers will not be disappointed at the lack of eureka moments or heroic problem-solving feats in the face of life-or-death consequences. Several of my chosen mathematical Americans did indeed live through highly dramatic experiences, but rarely, if ever, has the act of doing mathematics been central to the drama. The mathematical successes I re-count are more likely to involve methodical accumulation of knowledge than sudden brainstorms.

While the focus throughout is on individual lives, it has been part of my purpose to note how those lives have been affected by larger historical forces. I have tried to provide the reader with sketches of the political, mil-itary, economic, and social context needed to understand the choices faced by the people about whom I am writing. It should be no surprise to students of American history that developments that have repeatedly agitated the nation as a whole have also touched those involved with mathematics: com-mercial and industrial expansion, gender and race, war and the prepara-tion for war, immigration and sectional conflict. But this book does not claim to be a smooth narrative of 200 years of American history. There are many gaps, much backtracking within and between chapters, numerous pathways unexplored. My historical remarks are deliberately concise.

Naturally, I have also needed to say something about mathematical concepts. I hope that my mentions of basic topics in the school curriculum will be understandable to most readers: proportion, quadratic equations, the Pythagorean theorem. But I recognize that my brief attempts to con-vey something about more sophisticated ideas may be only faintly illumi-nating: the method of least squares, propositional logic, quaternions, the mean-value theorem, differential equations, non-Euclidean geometry, group theory, statistical mechanics, Fourier analysis. Readers are advised to gather what they can from these passages. A detailed understanding is not necessary for following the life stories.

I should make special mention of differential equations, a topic that sur-faces repeatedly in the following chapters. This occurred without particu-lar effort on my part, for, truth be told, compared to other topics in mathe-matics, I have never been that interested in differential equations. Writing this book has made me think I should once and for all admit this as a per-sonal failing and buff up my knowledge. As I note in the book's conclusion, however, some observers predict that the era of differential equations is on the verge of ending. It would serve me right to become an expert just as the subject becomes irrelevant.

Any biographical treatment of a given individual involves selection, all the more so for biographies as short as those in this book. In some cases, I know a lot more about the individuals than I have squeezed into the chapters. In a few cases, by choosing to emphasize other deeds and other anecdotes, I might have created in the mind of the reader a quite different picture. I will refrain from naming names here, other than to say that I am not referring to Abraham Lincoln, the subject of chapter 3. His many admirers, and his few detractors, will surely be little affected by whatever I say, or don't say, in this book.

The chapters can be read independently, but they do proceed essentially chronologically, and something may be gained by reading them in order. There are occasional cross-references. In the conclusion I highlight some of the trends revealed by the 200-year trajectory, as well as point out some ways in which my chosen few provide a misleading or insufficient picture of American mathematics. The Selected Bibliography, with a separate section for each of the other chapters, supplies the sources of all direct quotations and notes other significant supporting material used in writing the book.

The restriction to "Americans," meaning citizens of the United States, is convenient, but it should not for a moment be thought that American mathematical activity can be neatly separated from the rest of the world. As will soon became apparent, most of my mathematical Americans have had significant interactions with people or books from foreign lands.

═══════

A few facts about the state of mathematics at the time the United States became an independent nation in the late 1700s may help orient the reader who starts with chapter 1. There were nine colonial colleges, all of which have survived, although some under different names: Harvard, Yale, William and Mary, Dartmouth, Queen's College (now Rutgers), King's College (now Columbia), the College of Rhode Island (now Brown), the College of New Jersey (now Princeton), and the College of Philadelphia (now the University of Pennsylvania). These colleges, for men only, were largely intended to train a tiny minority of future lawyers, physicians, and clergymen. Some of the Founding Fathers were college men, some were not. John Adams graduated from Harvard and Thomas Jefferson attended William and Mary, but George Washington and Benjamin Franklin had no college training. All the colleges taught some mathematics, mainly arithmetic and geometry, but rarely gave it much emphasis. Nor was there much demand from the customers. Jefferson was unusual among students in his whole-hearted enjoyment of mathematics instruction, encouraging his professor to carry him as far as fluxions, the version of calculus propounded by Isaac Newton. No faculty member at these institutions was engaged in advancing mathematical knowledge. Such research activity would only begin to appear in the nineteenth century, and it would not be until the twentieth that it would become substantial.

Education for younger folk was a mixture of private tutoring and small local schools. Here again it was rare to find much focus on mathematics. In many places, especially in the South, only the aristocrats received much formal instruction of any kind on any subject. The strongest support for general education of the populace was in New England, but the emphasis was on learning to read the Bible. Basic arithmetic might be taught but was considered secondary. Textbooks were rare.

The education of girls was a low priority, varying widely, depending on individual circumstances. Abigail Smith, who later married John Adams, never attended any school. But her father, a Congregational minister and Harvard graduate, and her mother, daughter of another Harvard graduate, made sure that young Abigail could read and write and work with numbers. She even learned a bit of French. Moreover, Abigail was allowed to roam at will through her father's large private library, absorbing classics of English literature. With the literacy thus attained, the mature Abigail Adams expressed herself vividly on a host of subjects, including on the injustice of her limited education. After attending a public scientific lecture in 1787, she was moved to remark, "It was like going into a Beautiful Country, which I never saw before, a Country which our American Females are not permitted to visit or inspect . . . The Study of Household Good, as Milton terms it, is no doubt the peculiar province of the Female Character. Yet surely as rational Beings, our reason might with propriety receive the highest possible cultivation."

The education of children of color was a yet lower priority. As early as 1740, South Carolina had made it illegal to teach a slave to write. Nevertheless, striking instances of black literacy occurred in some of the colonies. Phillis Wheatley, after being brought in bondage from Africa to Boston as a child in 1761, mastered the English language in the course of religious instruction by her owners. She became a poet, acclaimed by George Washington, among others. Benjamin Banneker was born a free black on a farm in Maryland in 1731 and had little or no schooling. He somehow learned to read, possibly instructed by a Quaker neighbor. Eventually he taught himself enough algebra, geometry, and trigonometry to gain recognition as an astronomer and surveyor.

Astronomy and surveying were probably the most mathematically intensive activities in Colonial America. Many transactions for goods and services could be handled by barter or by the simplest of arithmetic calculations, with minimal concern for accuracy. But guiding an oceangoing ship by celestial navigation or laying out the boundaries of a tract of land were tasks that could be dramatically improved by precision. Some of the requisite mathematics was taught in the colleges, but more often this knowledge was transmitted not in a classroom but through private study and on the job. Thus did George Washington master enough mathematics to be a proficient surveyor, an important skill for a young man in a landholding family. Most navigators and surveyors were content with rules of thumb. Nathaniel Bowditch, the subject of chapter 1, was highly unusual in becoming deeply learned in the mathematics related to navigation.

At the end of the eighteenth century, mathematics in the newly independent country was mostly pursued for utilitarian reasons, but aesthetic appreciation was not unknown. John Adams, writing in 1783, recommended an algebra textbook to his son, John Quincy: "You will find it as entertaining as an Arabean Tale." And Thomas Jefferson, writing in 1799, lauded the practical utility of arithmetic, trigonometry, and a few basic propositions in Euclid's geometry. He declared quadratic equations and logarithms to be also of occasional help: "But all beyond these is but a luxury; a delicious luxury indeed; but not to be indulged in by one who is to have a profession to follow for his subsistence."

1

A Practical Navigator

Nathaniel Bowditch

1806

You will observe that I have carefully avoided all scientific display; I have written the entire work according to the current method of instruction in our country, where we prefer practical things to theory.

Nathaniel Bowditch

IN 1806 A RESIDENT OF SALEM, Massachusetts, Harvard class of 1777, remarked in his diary that "a self taught youth of this town, who has had no education or knowledge but in practical mathematics" had been appointed by the Harvard Corporation to the Hollis Professorship of Mathematics and Natural Philosophy. It is well to recall that in 1806 the United States, generously counting from the Declaration of Independence, was a mere 30 years old, while Harvard was 170 years old. Not so long compared with the venerable universities of Europe, but long enough for Harvard men to view themselves with no small esteem, to feel entitled to the envy of all outsiders, and to react with alarm at threats to the traditions of their insulated world. But the snobs need not have worried; the youth declined the offer. Probably his main concern was that a Harvard salary could not support his growing family. But in truth Harvard was intellectually beneath him.

The "youth" was Nathaniel Bowditch (1773–1838), and he was indeed self-taught, but his self-teaching was of a scale so extensive as has rarely been seen, before or since. Nor was he done learning in 1806. Bowditch eventually not only mastered the practical details of ocean navigation and coastal surveying but also acquired a comprehensive understanding of the most advanced European theories on the movements of the stars and planets.

Bowditch parlayed his mathematical skills to achieve a comfortable career in insurance and banking. It was not so much that he performed elaborate calculations related to these fields, but rather that his status as a calculating wizard gave him an aura of probity and objectivity that served to make his reputation as a preeminently trustworthy man of business. In part he was the epitome of the American achiever of the early nineteenth century, rising from humble forebears who had made their living primarily through manual labor to become a man of substantial social and financial standing. He was more distinctive in becoming both a self-made man of business and a self-made scholar.

Mathematics today in the United States is largely encountered as a series of strictly ordered credential hurdles in school: arithmetic, algebra I, algebra II, . . . , on to undergraduate and graduate degrees in mathematics or related disciplines. In Bowditch's day such regimented credentialing was only dimly taking form, and he largely sidestepped even such as then existed. He was known in his later years as Dr. Bowditch, on the basis of an honorary doctor of laws degree conferred by Harvard in 1816, although he had never attended that or any other educational institution after the age of ten. And although Harvard would make little impression on Bowditch, he would ultimately leave a substantial mark on Harvard.

Salem, a seaport about 15 miles northeast of Boston, was older than Harvard, tracing its origin to 1626. It had famously been the site of the witch hysteria of 1692, a date only just passing from living memory when Nathaniel Bowditch was born. In Bowditch's lifetime a new source of fame for Salem would rise to a peak and then begin to decline: the era of maritime trade with the East Indies. Beginning soon after the Revolution, Salem became a leading center for commerce with Asia, regularly sending ships to India, to the islands of what are now the Philippines and Indonesia, and even to China. A particular specialty was the pepper trade with Sumatra. In 1790 Salem was the sixth-largest city in the United States. As late as the 1830s one Asian potentate believed that Salem was an independent country, and one of the richest in the world.

Asia-bound ships out of Salem generally took the eastern route, working down the west coast of Africa before rounding the Cape of Good Hope, sometimes making trading stops on the east coast before striking out for India or the spice islands of Southeast Asia. With the only motive power being the wind in the sails, these were long voyages, often lasting a year or more, but the rewards could be magnificent. The also lucrative slave trade was only a minor sideline for the Salem shipping industry, certainly as compared with the large fortunes piled up through slave commerce by leading citizens of Newport and Bristol in neighboring Rhode Island.

Massachusetts officially prohibited participation in the slave trade in 1788. The owning of slaves had been gradually declining there for some years, although the exact date of abolition of the institution is unclear. About 6,000 lived in bondage in the state in 1790. Bowditch's view was that slavery was "one of the greatest of moral evils," but that blacks were "a race of men naturally less intelligent than the whites." He advocated gradual emancipation, with compensation to the slaveholders. After Bowditch's death, his youngest son, under the influence of Frederick Douglass and William Lloyd Garrison, took a more radical abolitionist line, publishing in 1849 an erudite and angry book that demonstrated, and lamented, that slavery was fundamentally embedded in the Constitution of the United States. Young Bowditch lambasted the slave owners, and all who gave aid and comfort to them, North and South, as responsible for whatever mental and moral shortcomings might be observed among the enslaved.

═══

Nathaniel Bowditch's forebears had settled in Salem from England before 1640, with the men all making a living by venturing on the ocean, usually supplemented with some land-based trade of importance to the nautical economy. His father was a cooper, making and repairing barrels.

Young Nathaniel attended a school in Salem, but family finances required that he leave school at the age of ten, first assisting his father in the barrel trade, and then working in shops provisioning the ships of Salem for their long voyages. In the intervals between customers he read avidly, becoming intimately familiar with Shakespeare and the Bible, but giving special attention to mathematics. Before he was able to afford to buy books,

Bowditch frequently borrowed them from the Salem Athenaeum, which was unusually well supplied with scientific books, owing to a peculiar circumstance of the Revolutionary War. In 1780 an American privateer captured a ship transporting a large private library from Ireland to England, including many volumes of a scientific or mathematical flavor. The privateer's home port was Beverly, adjacent to Salem, and according to Bowditch's son, writing in 1838, the following ensued: "The enlightened and liberal owners of the vessel permitted the library thus captured to be sold at a very low rate to an association of gentlemen in Salem, and it became the basis of the present Salem Athenaeum." Similar events in connection with more recent armed conflicts have not been much in evidence.

Bowditch made five long ocean voyages between 1795 and 1804, advancing in rank from clerk, to supercargo (the representative of the owners of the ship and of the cargo), and finally, on his last voyage, to the full command as shipmaster. He visited ports in Spain and Portugal; he visited the Isle of Bourbon (now Réunion) in the Indian Ocean, east of Madagascar; and in Asia he visited Batavia, Manilla, and Sumatra.

These voyages included long stretches of time when he had little or no responsibilities. Bowditch used this time to improve his knowledge of both mathematics and languages. He had already taught himself Latin in the 1790s, to read Newton's *Principia Mathematica*, and by the end of his last voyage he could read French, Italian, Portuguese, and Spanish. Later, in his forties, he learned German.

Long-distance sailing in the eighteenth century relied on a variety of technical principles that were being helpfully compiled into books for ease of use by seamen. By the 1790s the principal volume on these lines for the English-speaking world was that of London's John Hamilton Moore (1738–1807): *The New Practical Navigator*. An unauthorized American edition was being produced in Newburyport, Massachusetts, not far from Salem. Copyright in the early United States was only casually observed, especially for books originating in the country from which the United States had so recently emancipated itself. When the Newburyport publisher sought to respond to increasing complaints about errors in the text, he turned to the local calculating whiz, Bowditch. Bowditch would make so many changes (over 8,000, he would claim), and his fame among sailors would become so significant, that in 1802 the publisher took Moore's name off and put Bowditch's name on, while altering the title to *New American Practical Navigator*. And so it has remained, down to the present day.

Bowditch's book, like Moore's before him, guides the sailor in determining where he is on the surface of the earth as measured by two numbers—latitude and longitude—and how to proceed to somewhere else. The primary astronomical aids to these tasks are the positions of the sun, moon, planets, and stars. For this reason a good deal of attention in the book is given to instruction in using sextants and other instruments to make observations of these heavenly bodies.

The primary mathematical tool discussed by Bowditch in the *Navigator* is trigonometry: given knowledge of some of the side lengths and angle measurements of a triangle, find the measures of all the other sides and angles. This is exactly what is needed for calculating the straight-line distance of a ship from its starting point, given that it has sailed a certain distance on one heading, and then a further distance on another heading. All other navigational exercises are elaborations of this simple case. Of course, only for short distances is it safe to proceed as if the earth is a flat plane; hence the term "plane sailing," also the title of a section of Bowditch's book, where the calculations are as elementary as possible. For longer distances, the fact that the earth is a sphere must be taken into account, and Bowditch diligently explains the appropriate modifications needed.

Bowditch was more comprehensive than Moore in discussing the variety of methods for determining longitude, a delicate business that perplexed sailors for centuries. It takes only modest observational skill to notice that a change in latitude—traveling directly north or directly south—affects the number of hours of daylight. Thus most readers will be willing to believe that latitude can be determined by some procedure involving measuring the maximum height of the sun above the horizon.

Detecting a change of longitude is more complicated, since traveling directly east or west has no effect on the observed path of the sun across the sky. The well-traveled reader may have noted that journeying east or west most definitely does affect the time at which the sun rises or sets.

Surely one could do something with this fact to measure the longitude. Exactly so. As Bowditch himself expressed it in his *Practical Navigator*, "if a clock or watch could be so contrived, as to go uniformly in all seasons, and in all places, the longitude might easily be deduced therefrom." The difficulty, so hard to appreciate in a world saturated with accurate timekeeping mechanisms, is in contriving the accurate clock or watch. It was not until the 1760s that Englishman John Harrison conclusively demonstrated a suitable chronometer, able to reliably withstand conditions at sea. This breakthrough was celebrated in Dava Sobel's 1995 book *Longitude*. But even in editions of Bowditch's *Navigator* in the 1820s, he still devoted much attention to explaining complex alternative longitude methods, such as measuring the angle between the moon and other heavenly bodies (called the method of lunars), or observing eclipses of the moons of Jupiter. It is thus evident that the Harrison technology revolution took a long time to triumph. Part of the delay arose from the high cost of truly reliable chronometers. But another part can be attributed to the pleasure taken by mathematicians such as Bowditch in contemplating and in carrying out more delicate, theory-laden measurements such as the method of lunars.

It was Bowditch as the author of the *Practical Navigator* that Harvard attempted to hire in 1806, a man whose eventual far-flung fame was not yet clear, and so whose seemingly limited intellectual range could be lampooned. But Harvard at the time had its own distinctly limited intellectual range. In 1800 it was graduating only about 40 men a year. The faculty consisted of the president, three professors, and four tutors, and the college was still under clerical control, although the proportion of graduates going into the ministry had declined from about 40% early in the eighteenth century, to about 20% as the century turned. The relative triviality of the mathematics instruction at Harvard, concentrating on the basics of arithmetic, algebra, and geometry, would have provided limited inducement for Bowditch.

By the time of the Harvard offer, Bowditch had retired from the sea and established himself as an executive in a Salem banking and insurance firm, the Essex Fire and Marine Company, far more lucrative than a Harvard professorship. He left seafaring at just the right time. The Jefferson and Madison administrations, attempting to pilot a course between the warring

Engraving of Nathaniel Bowditch, by G. F. Storm after Gilbert Stuart, ca. 1828. Harvard Art Museums / Fogg Museum, Gift of the Estate of Miss Williams through Henry M. Williams, M500. From the Imaging Department © President and Fellows of Harvard College

nations of England and France, made a series of policy decisions, most notably the trade embargo of 1807 and the War of 1812, that devastated US shipping interests. The eventual national recovery, though robust, largely bypassed Salem in favor of other ports. Another famous son of Salem, the writer Nathaniel Hawthorne, described his home town in the 1840s as "scorned . . . by her own merchants and ship-owners, who permit her wharves to crumble to ruin, while their ventures go to swell, needlessly and imperceptibly, the mighty flood of commerce at New York or Boston." Gone forever were the great days of Bowditch's early manhood, "when India was a new region, and only Salem knew the way thither."

Bowditch himself would later follow the winds of change buffeting Salem, leaving his native town for Boston in 1823 to become the chief executive of the Massachusetts Hospital Life Insurance Company, for the handsome salary of $5,000 per year, more than triple his current compensation in Salem. In the intervening years he had declined offers of professorships from the University of Virginia (despite a direct appeal from Thomas

Jefferson) and from West Point (turning aside a request from Secretary of War John C. Calhoun).

As during his sailing days, Bowditch's business responsibilities seem to have given him ample time to indulge his intellectual passions. He became a regular contributor to mathematical and astronomy journals, both foreign and domestic. His erudition and his humility quickly won him the respect of European experts and would lead to honorary membership in the Royal Society of London and other foreign scientific organizations. One of his frequent correspondents, the peripatetic middle-European astronomer Franz Xaver von Zach, lauded Bowditch as "the first, and up to the present the only, great geometer in America."

While he was still living in Salem, Bowditch conceived, and largely completed, a massive project that fully taxed all his knowledge of mathematics and astronomy: a translation of the *Méchanique Céleste* by Pierre-Simon de Laplace. Isaac Newton, in his celebrated *Principia* of 1687, had been able to show how fundamental features of planetary and lunar motion, as well as the behavior of projectiles on the surface of the earth, could be explained by some simple laws governing the behavior of material objects. Prime among these laws was universal gravitation, which asserted that each piece of matter in the universe attracted every other piece of matter in a rigorously specified manner. Starting in 1799, Laplace had been engaged in extending and enriching the Newtonian program, bringing to bear a host of helpful mathematical developments that had emerged in the intervening century, as well as new observational data. If one knows the position and velocity of an object at a particular time, and one knows the forces acting on the object, it may be possible to predict where the object will be at all future times, utilizing the fundamental mathematical tool known as a differential equation. The study of these equations had arisen out of the differential and integral calculus pioneered by Newton.

When Bowditch began his translation project, Laplace had produced four volumes of his work. Bowditch completed his translation by 1817 but did not publish it until 1829. (A fifth volume, published by Laplace in 1825, never received the full Bowditch treatment.) Bowditch rejected offers of financial assistance to publish, saying that he did not wish to impose such a burden on those who would not be able to read and benefit from the work itself.

Bowditch set himself the task of rectifying what he found to be Laplace's overly concise exposition and failure to properly acknowledge the work of his predecessors. He also noted outright errors on the part of the Frenchman. Bowditch helped substantially to fix Laplace's reputation as an aloof, arrogant know-it-all. Biographical accounts of Laplace frequently mention that his *Méchanique Céleste* blithely leaps over conceptual gaps and delicate derivations with phrases such as "it is easy to see," a stylistic habit noted by Bowditch, who complained that when he came across such Laplacian rhetoric he knew he had hours of work ahead of him. For example, on page 6 of volume 1, Laplace's "easy to infer" gives rise to a one-page Bowditch footnote. More than 50% of Bowditch's rendition of Laplace's chapter 1 is composed of Bowditch's footnotes.

Bowditch's translation of Laplace was a source of pride to his fellow Americans, but the praise of European readers often contained an element of condescension, suggesting that Bowditch was mainly remarkable for surpassing their low expectations for American intellectual achievement. It was commendable for Bowditch to have struggled through Laplace, but we Europeans, it was implied, had meanwhile moved on to new frontiers. In part Bowditch shared this view of what he had done. He mused on the difference between two famous mathematical figures of the ancient world: Euclid and Archimedes. Most historians of mathematics would agree, then and now, that Euclid's far-famed book on geometry was a compilation of the ideas of other, more original, thinkers. Archimedes, in contrast, was the real article, a towering mathematical genius. And so Bowditch compared himself to Euclid, while likening Laplace to Archimedes.

Bowditch saw Laplace as standing head and shoulders above his contemporaries—a second Newton. History has judged this assessment as perhaps slightly exaggerated, but a more serious misapprehension on the part of Bowditch was his failure to see how much of Laplace's success was built upon a community of scholars. The French education system was already showing signs of crystallizing into the modern form, with research universities at the top, staffed by elite researchers like Laplace, and a range of lesser institutions, and less brilliant thinkers, crucial to making the whole thing function. It never seems to have occurred to Bowditch that he might have used his brainpower to create such a community in the United States. In future chapters we will see, at West Point and the *Nautical Almanac* Office, men who were distinctly less gifted than Bowditch nevertheless make a deeper mark than he, by employing their skills in a communal setting.

As time passed, Harvard took a larger place in Bowditch's life. Three of his four sons graduated from the college, and in 1826 Bowditch joined the Harvard Corporation as one of a small group of local business leaders who, together with the college's president and treasurer, managed the day-to-day operation of the institution. Bowditch moved aggressively and effectively to repair the financial health of the college, with little deference to established customs or personal feelings. Reports that he was responsible for forcing the resignation of the well-loved president, John T. Kirkland, created ill will in Cambridge for many years after Bowditch's death in 1838.

Despite expending more energy in making Harvard a business success than in making it a center of scholarship, Bowditch ultimately helped launch a mathematical research tradition in the United States, by his discovery and encouragement of another Salem boy, Benjamin Peirce (1809-80). Peirce would likely have attended Harvard anyway; his father had been in the class of 1801. But it is doubtful that Peirce would have been so early exposed to higher mathematics without Bowditch's intervention. Peirce attended a grammar school in Salem with one of Bowditch's sons, who alerted his father to his schoolmate's budding mathematical gifts. When Peirce was a Harvard undergraduate in the late 1820s, Bowditch hired him to assist in proofreading the Laplace translation then being readied for publication. Peirce would eventually be appointed as a professor of astronomy and mathematics at Harvard. Unlike Bowditch, Peirce did not decline the offer, nor did he hesitate in attempting original research, and in joining and forming intellectual teams. Peirce came to be recognized as a major national figure in science and mathematics in the United States, a status to which Nathaniel Bowditch, for all his talent, never seems to have truly aspired.

2

Hudson River School

Sylvanus Thayer

Like the house that defied the storm, West Point is built on a rock, and
that rock is mathematics.

C. De W. Willcox

I N THE SPRING OF 1815, with the US military returning to peacetime
status after the conclusion of the War of 1812 with Great Britain, a two-
man team was sent to Europe on behalf of the Military Academy at West
Point, New York. They had been selected for this mission, as "two of our
best officers," by the army's chief of engineers. The choice was ratified by
Secretary of War James Monroe, after a personal interview with the duo in
Washington, DC. The junior officer of the team was Sylvanus Thayer, a
thirty-year-old brevet major, which is to say that he had been nominally
promoted for meritorious service, without an increase in pay. The goal of
the trip was to improve the academy by learning about European methods
for educating apprentice military officers, and to acquire whatever books,
maps, and other objects might be deemed relevant. While the two men
were inspecting facilities in Paris, Metz, and other locations, the Military
Academy was thrown into disarray by conflict between the superintendent, Alden Partridge, and the faculty. The situation had reached such a
crisis in June 1817 that Monroe, who was now president of the United
States, determined to remove Partridge and to name a new superintendent. He chose Major Thayer, who took command of the academy immediately upon his return from Europe in July 1817. Never again has the
academy had a superintendent so young, or so influential.

≡≡≡

Thayer, the son of a Massachusetts farmer, had himself graduated from West Point in 1808, with less than a year of residence, benefiting from the lax requirements at that time. But he was already a man of some educational attainment, having earned a bachelor's degree at the top of his class from Dartmouth College just the year before. Dartmouth, although more than 100 years younger than Harvard, was nevertheless able to boast of being one of only nine colleges in existence before the Revolution. Thayer would teach some mathematics at the academy, before being assigned to assist with coastal fortifications during the War of 1812.

The Military Academy had been muddling along since its official founding in 1802, during the first administration of Thomas Jefferson. Not many years before, Jefferson had denounced the very idea of a military academy, but as president he discovered that it had political advantages, allowing him to stock the military with young men loyal to his own Republican party, opposed to the Federalists. Jefferson did not expend much effort in choosing either the location or the curricular focus of the academy, essentially just allowing previous suggestions on these matters to take the path of least resistance. West Point had had troops and fortifications ever since the Revolutionary War, during which its strategic importance on the Hudson River became evident, with traitor Benedict Arnold infamously seeking to deliver it into British control. When Congress created a Corps of Artillerists and Engineers in 1794, it was headquartered at West Point, and it was this technically oriented corps that was most in the minds of those agitating for a military academy. It seemed obvious that those charged with launching projectiles and building fortifications could benefit from some classroom learning.

Military engineers, in eighteenth-century Europe and its colonies, were the intellectuals of the battlefield. They were the behind-the-scenes technical experts who measured the distances and designed the fortifications, who pondered the angles of impinging forces and the heights of the walls, the ones who defined the environment in which others did the advancing and retreating, the bombarding and excavating, the killing and dying.

Military engineering, like mathematics, had developed a specialized lingo that could be wielded to mystify those not in the know. In the 1760s Laurence Sterne had satirized the field in his mischievous romp of a novel, *The Life and Opinions of Tristram Shandy*. The title character's Uncle Toby is obsessed with the fortifications used in the Anglo-French wars waged between 1689 and 1713. He invokes the great names of the military engineering literature, such as Galileo, Tartaglia, and Vauban, all of whom would become well known to the students at nineteenth-century West Point. Toby perplexes the other characters with his attempts to explain

> the differences and distinctions between the scarp and counterscarp,—the glacis and covered way,—the half-moon and ravelin . . . For when a ravelin, brother, stands before the curtin, it is a ravelin; and when a ravelin stands before a bastion, then the ravelin is not a ravelin;—it is a half-moon;—a half-moon likewise is a half-moon, and no more, so long as it stands before its bastion—but was it to change place, and get before the curtin—'twould be no longer a half-moon; a half-moon, in that case, is not a half-moon;—'tis no more than a ravelin.

This is not far removed from the historical survey of fortification theory provided by a Military Academy textbook: "The covered-way was introduced, and became an integral part of the front; and a small demilune, or ravelin was placed in advance of the enceinte ditch, forming a tête de pont to cover the communication, at the middle of the curtain across the main ditch, between the enceinte and the exterior."

The author of these words was Dennis Mahan, a West Point student under Thayer, later appointed to a professorship of engineering under Thayer's regime. "Much that appertains to the Engineers Art," Mahan asserted, "is but an affair of feet and inches." The great goal of fortification was to fire on the enemy while remaining as nearly invisible as possible to that enemy. From this followed an emphasis on measurements of lengths and angles. No good military engineer could do without a thorough grounding in geometry and trigonometry.

It was thus quite natural that West Point should emphasize mathematics from the start, much more so than the established colleges, like

Dartmouth, Harvard, and Yale, which were largely still content, early in the nineteenth century, to order the curriculum around the needs of future clergymen. But it required vison and hard work to create the conditions whereby mathematics at West Point could be made part of a strong institutional structure.

On his European sojourn Sylvanus Thayer had observed an impressive model for structuring technical education: the École Polytechnique in Paris, founded in 1794 by the Revolutionary government. Mathematician Gaspard Monge, one of the initiators, became an intimate friend of Napoléon Bonaparte, himself an accomplished student of mathematics. Rarely, if ever, has there been such a close and long-lasting relationship between a first-class mathematician and a powerful political leader. Monge had pioneered the field of descriptive geometry well before the French Revolution. Monge had essentially made systematic what artists such as Albrecht Dürer had been doing informally for some while: the depiction of three-dimensional objects through two-dimensional horizontal and vertical projections. Having produced this work specifically for designing fortifications and cannons, Monge's work on descriptive geometry was initially considered too militarily sensitive to publish. But after the Revolution he made it a staple of the curriculum at the École Polytechnique. Likewise, it became a staple at West Point.

Certain important aspects of the École Polytechnique in its first decades deviated from the original intentions for the institution and escaped the control of both Monge and Napoléon in ways that would foretell developments in the United States later in the nineteenth century. The school was supposed to provide general preparation so that graduates could then proceed to more advanced study at several specialized "application schools." In practice, however, prestige and intellectual eminence accumulated at the École Polytechnique, with the application schools being staffed largely by lesser lights. The engineering emphasis championed by Monge was gradually diminished by the force of a set of brilliant mathematicians, many of whose names have resounded down the subsequent decades of mathematical history, with J. L. Lagrange and P. S. Laplace perhaps the most illustrious. Having mastered the seventeenth-century calculus of

Newton and Leibnitz, and the century of subsequent elaboration of this mathematical milestone, they authored such towering masterpieces as *Mécanique analytique* and *Traité de mécanique céleste*, and promoted abstract mathematics at the École Polytechnique far beyond the needs of engineers.

When Thayer arrived in Paris, he found Napoleon at last decisively overthrown, and Monge dismissed. In fact, the entire École Polytechnique had been closed for being a hotbed of radicalism. It soon reopened, however, with perhaps even more emphasis on mathematics than before. Laplace wanted to rename it "École Mathématique."

But above all Thayer would find in the Polytechnique the great exemplar of a trend in the institutionalization of advanced mathematics that would continue to grow through the nineteenth century and into the twentieth: mathematicians as educators. Whereas institutions such as Frederick the Great's Berlin Academy, the home of Lagrange for many years before he came to the École Polytechnique, had employed mathematicians to glorify the sovereign and sometimes to provide technical advice to the government, the Polytechnique explicitly gave mathematicians the mission of teaching and examining the new generation, thus exerting an influence beyond the achievements of any one person, or any one generation of scholars. West Point would not become another École Polytechnique, and it would not employ anyone as accomplished as Lagrange, Laplace, or Monge, but it too would have a multiplier effect, in its more modest way, by bringing together a critical mass of mathematically knowledgeable individuals and giving them a teaching mission.

In his early years as superintendent, Thayer managed to regularize certain features of the Military Academy that will seem to the modern reader of such obvious importance that the previous failure to fully address them may be surprising: adequate housing for faculty and students, daily schedules for classes and other activities, yearly examinations for class advancement, and a strict policy of admitting students only at one set time each year; previously, students had been allowed to wander in at their own convenience. Thayer also insisted on small classes, a practice followed by the École Polytechnique that became firmly fixed at West Point. His high

standards for student conduct were not always popular, but this too established a long-lasting tradition. Thayer even weathered the embarrassing
and uncomfortable fact that his predecessor as superintendent, Alden Partridge, had initially refused to accept that Thayer had replaced him. Although a court-martial procedure eventually extracted Partridge from the
academy with finality, he continued to be an aggravating critic of Thayer
for many years.

The mathematics and engineering faculty, and the students who graduated during Thayer's tenure at West Point, provided mathematical influence locally and nationally that would extend well beyond his own departure as superintendent in 1833. Professor of Engineering Claude Crozet, a
student of Monge at the École Polytechnique, would introduce descriptive
geometry to the academy and publish his own textbook on the topic.
Charles Davies would publish eight mathematics textbooks as professor of
mathematics at West Point from 1823 until 1837. He would go on to write
41 additional titles, from basic arithmetic to calculus, selling over seven
million total volumes by 1875, standing atop the mathematics textbook
market in the United States through much of the nineteenth century. Albert Church, after graduating from West Point in 1828, taught mathematics
there for 48 years. He too wrote mathematics textbooks, though less successfully than Davies.

In keeping with Thayer's admiration for the École Polytechnique,
French textbooks were heavily relied on during his regime at West Point,
and consequently the French language was an important part of the curriculum. French continued to be emphasized even after Thayer's departure,
as testified to by both William T. Sherman (class of 1840) and Ulysses S.
Grant (class of 1843) in their memoirs, even though French textbooks were
being replaced by English language books, such as the Davies mathematics
series. Several of Davies's books were indebted to French models, and in
some cases they were simply translations.

The events of the Mexican War and the Civil War clearly demonstrated to observers far and wide that West Point was producing accomplished military leaders with regularity. During this same period the
school also became, with less publicity, the nation's most important train-

ing ground for civil engineering. By the 1820s there was widely voiced demand for transportation improvements to support the growth of the commercially minded young country. In response, pressure was exerted from political leaders in Washington, notably by Secretary of War John C. Calhoun, to enlarge the academy curriculum so that graduates would build not just forts and intrenchments, but also roads and bridges and canals. Thayer, left to his own inclinations, would not have chosen this path, but it proved to be one of his legacies. His appointees as professors of engineering, first David Douglass and then Dennis Hart Mahan, were vigorous in carrying out the civil engineering mandate. Mahan, of the class of 1824, was a faculty mainstay for over 40 years. He was firm in defending the institution against outside critics, and he published books on fortification, civil engineering, and industrial drawing. This last aimed at bringing the elementary principles of descriptive geometry to the secondary schools.

Least well recognized of all, West Point before the Civil War quietly became a significant training ground for teachers of mathematics nationwide. At least 41 West Point graduates or former faculty members taught this subject at colleges or universities other than West Point itself prior to the war. Looking at it another way, of the 203 colleges and universities that existed in the United States in 1860, at least 37 (more than 18%) had had one or more mathematics professors who were either Military Academy graduates or had taught at West Point. And these numbers do not include such distinguished graduates as Robert E. Lee, who briefly taught mathematics at West Point, or Ulysses S. Grant, who wished to teach college mathematics but whose life circumstances prevented it. In addition, it is likely that many academy graduates taught mathematics in secondary schools.

Thayer also influenced other institutions of higher learning in more indirect ways. George Ticknor, an old Dartmouth classmate of Thayer who had remained a close friend, became a professor of French, Spanish, and belles-lettres at Harvard. When he began to agitate for reform at Harvard in the 1820s, he pointed to Thayer's example at the Military Academy. In 1826 Thayer's one-time champion, James Monroe, now a former

Brigadier General Sylvanus Thayer, by Robert W. Weir, 1843. Part of the US Military Academy class of 1808, Thayer was US Military Academy superintendent from 1817 to 1833. Courtesy of the West Point Museum, US Military Academy

US president and a regent of the University of Virginia, turned to Thayer for advice on governing that institution, established by Thomas Jefferson in 1817.

When Andrew Jackson, self-styled representative of the common man, took over the presidency of the United States in 1829, he looked on West Point with some suspicion as a haven for the privileged. He also scoffed at the school's disciplinary policies, often overruling Thayer's attempts to dismiss disobedient cadets. Thayer, increasingly distressed by the evident disrespect from Washington, finally offered his resignation, which was accepted in 1833. He would remain on active duty as an army engineer until retiring as a brevet brigadier general in 1863. He was especially noted for designing the harbor defenses around Boston. Before he died, in 1872, he gave a substantial endowment to his alma mater, Dartmouth, to establish a school of engineering, which was named in his honor.

Sylvanus Thayer had been superintendent of the Military Academy for sixteen years. No other superintendent has served more than eight years; most have served much less. Consequently, his impact lingered long. It was not until after the Civil War that Thayer's students began to relinquish control of the West Point faculty. The year 1871 marks a transition date, when professor of natural and experimental philosophy William H. C. Bartlett (class of 1826) retired, and professor of engineering Dennis Hart Mahan committed suicide by jumping from a steamer on the Hudson River. This latter event shocked and saddened the academy, but it did not impede one late instance of Thayer's influence. Mahan's son would become the greatest theorist of naval power of his time, read avidly by Theodore Roosevelt and by German and Japanese strategists nurturing imperial ambitions. This naval advocate, named for his father's mentor, was Alfred Thayer Mahan.

3

Political Arithmetic

Abraham Lincoln

1826

But what, at last, is this proposition? I believe it is a sort of proposition in proportion, which may be stated thus: As the negro is to the white man, so is the crocodile to the negro, and as the negro may rightfully treat the crocodile as a beast or a reptile, so the white man may rightfully treat the negro as a beast or a reptile. That is really the "knip" of all that argument of his.

Abraham Lincoln

SOMETIME IN 1826 A LANKY teenager in rural Indiana carefully wrote out an arithmetic problem on a sheet of paper, using homemade ink and a quill made from the feather of a turkey vulture. This sheet of paper was sewn up as the last entry in a small booklet with similar sheets of arithmetic problems. Within a few years the young man would leave Indiana for more populous locales, where he would eventually achieve great fame, before dying suddenly at the age of fifty-six. His stepmother preserved the arithmetic booklet, and when she was approached by her stepson's former law partner, collecting biographical information not long after his partner's death, she gave this lawyer the book. Over the next two decades the lawyer disbursed the pages, at first in the spirit of generosity to admirers of his famous partner seeking autographs, but in later years he sought payment, to relieve his financial distress. The pages remain disbursed to the present day, but scholars have succeeded in identifying the location of 22 of them, preserved in libraries and private collections.

The teenage arithmetician was Abraham Lincoln (1809–65), and the booklet was his "cyphering book." It recorded his efforts, between 1819 and 1826, to learn arithmetic, from the basic operations of adding, subtracting, multiplying, and dividing, up to more advanced topics, such as the "double rule of three" and compound interest. These handmade booklets were a central tool of mathematical instruction for many Americans until the middle of the nineteenth century, when printed textbooks became more widely accessible. A cyphering book was the creation of an individual student, who would painstakingly copy mathematical rules and problems from the dictation of a teacher, who might be reading from an intact textbook, another cyphering book, or from some hodgepodge of pages collected by the teacher over a period of years. Lincoln's cyphering book reflects the influence of at least four different published arithmetic textbooks.

Lincoln's school days were spent in rural Kentucky and Indiana, where teachers of even the lowest quality were a rare sight. They, and the population they served, were restless, migrating frequently in search of better opportunities. Lincoln encountered at least six different teachers, none for more than a few months at a time. Only during winter could his family spare him from the hard labor of scratching a living out of the land. All of Lincoln's schools were one-man, one-room operations, serving both boys and girls of a mix of ages. The teacher was usually paid in farm produce or animal pelts. Most of Lincoln's cyphering book was likely written within one or more of these schoolhouses, since only in such a place could he find a flat desk for ease of writing. Lincoln's home at this time, as befits his legend, was a log cabin, a mere 20 feet by 18 feet, in which he was only one of as many as eight residents. Nevertheless, acquaintances of his youth testify that he would practice his arithmetic in these crowded conditions, scrawling figures on a board or wooden shovel with a charred stick by the light of the fireplace.

In later years Lincoln enjoyed emphasizing his lack of formal education. In 1859, as his name began to be put about as a presidential candidate, he included the following description of his schooling to the secretary of the Illinois Republican state central committee:

There were some schools, so called; but no qualification was ever required of a teacher, beyond *"readin, writin, and cipherin,"* to the Rule of Three. If a straggler supposed to understand latin, happened to sojourn in the neighborhood, he was looked upon as a wizzard. There was absolutely nothing to excite ambition for education. Of course when I came of age I did not know much. Still somehow, I could read, write, and cipher to the Rule of Three; but that was all.

The "rule of three" that Lincoln mentions was the phrase used in virtually all arithmetic books of that era to describe the method of solving proportion problems, in which an unknown quantity was to be deduced from three given quantities. For example, Lincoln wrote in his cyphering book, "If 3 Oz. of silver cost 17s what will 48 Oz. cost," and proceeded to multiply 17 by 48 and then divided by 3. In doing this he was following a "rule" he had recorded on the previous page of his cyphering book, with quirky spelling:

How is the fourth term found in
[dire]ct proportion
[b]y multipliing the 2nd and 3rd
together and dividing that
by the first term

He completed this particular problem by dividing by 20 to convert from shillings to pounds. That English money was still present in the arithmetic instruction of this young American was not unusual. It would be another couple of decades after Lincoln's boyhood before the English influence was diluted enough that exercises using federal money became the norm.

Proportion remains an important topic in elementary mathematics, but it is now usually subsumed within beginning algebra. The proportional relationship in the silver problem can be translated as an equation of fractions in several different ways. For example:

$$\frac{3}{17} = \frac{48}{x}.$$

This can then be solved for the unknown x, by the general rules for solving algebraic equations, which the student today is encouraged to regard as more fundamental than the specific manipulations needed to solve proportion problems. Like most of his contemporaries, Lincoln remained ignorant of algebra; it was only toward the close of the nineteenth century that there was agitation to make it a standard school subject. There can be little doubt that the student who adopts the algebraic approach to proportion problems is conforming more closely to the main line of development of the field of mathematics, stretching upward to the research universities. Whether it makes the student more likely to solve a given proportion problem, however, may still be subject to debate.

In claiming that the rule of three marked the limit of his youthful training in mathematics, Lincoln was advertising himself as a regular fellow whose school learning would be considered comfortably modest by a wide circle of the general public. Lincoln, as has long been remarked, was an image-conscious politician. "His ambition," according to William Herndon, the law partner who once owned the cyphering book, "was a little engine that knew no rest." In fact Lincoln's arithmetic education went further than he acknowledged; not a lot, but enough that it was not so "defective," as he preferred to style it. The surviving pages of Lincoln's cyphering book show him working on topics consistently treated after the rule of three in the arithmetic instruction of the time: double rule of three (setting up proportions to find an unknown quantity from five known quantities), simple and compound interest, and discount.

Some of Lincoln's most important political contemporaries received significantly more schooling than he did. James Buchanan, Lincoln's predecessor as president, had graduated from Dickinson College in Pennsylvania. Stephen Douglas, Lincoln's adversary for the Senate and later the presidency, though not himself a college graduate, was the son of a college graduate and had attended a substantial Vermont secondary school that has persisted to the present day, unlike the ephemeral schools encountered by Lincoln. Jefferson Davis, Lincoln's antagonist as president of the Confederate States of America, was a West Point graduate. William Seward, the one-time rival whom Lincoln appointed as his secretary of state, had a

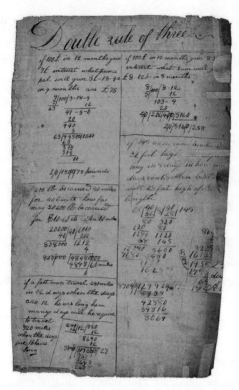

Page from Abraham Lincoln's mathematical exercise book, ca. 1825. The page was gifted to Harvard by Christian A. Zabriskie in 1954. MS Am 1326, Houghton Library, Harvard University

degree from Union College in New York. But such men represented only a thin crust of American society. The most careful recent research into Lincoln's education concludes that he attended school for about 12 months, stretched out over several years, and that this was in fact "above the average for children in the United States, especially for children in the midwest."

Moreover, there was a large population in the United States whose members would have looked with envy at the educational opportunities available to Lincoln. Many states had laws on the books explicitly prohibiting the educating of slaves, and some, like South Carolina, extended this prohibition to free people of color. Even where no such statutes existed, public attitudes could effectively prevent schooling of African Americans. Thus in 1827, in Baltimore, Maryland, the owner of ten-year-old slave Frederick Douglass quashed his wife's efforts to teach the youngster to read by declaring that "A nigger should know nothing but to obey his master—to do as he is told to do. Learning would *spoil* the best nigger." Nor were conditions necessarily better in the North. In the 1830s Prudence Crandall of Canterbury, Connecticut, was forced to abandon her attempts to educate young women of color after being subjected to repeated legal roadblocks and physical violence.

Lincoln once declared that he had had personal experience of life under slavery, citing the hard life of subsistence farming he lived in Indiana, especially the time that his father rented him out as a general laborer: "I used to be a slave." This must be judged an exaggeration.

But if we compare Lincoln's environment with that of the autodidact of chapter 1, Nathaniel Bowditch, it is surely fair to say that Lincoln's efforts at self-education faced far greater challenges, with Bowditch's world relatively awash in learning, both abstract and practical. For the young Lincoln there was no Harvard down the road, and no town in Illinois, Indiana, or Kentucky in which he lived had a substantial library of books. Bowditch's sea voyages also gave him far more experience of the wider world than Lincoln had, and a greater opportunity to closely observe the technical details of business and commerce. In only one realm did the young Lincoln have any comparable practical experience: surveying.

Surveying, along with a legal system for adjudicating property rights, undergirded the westward march of the land-hungry multitudes in the nineteenth century. Vast new tracts of "unoccupied" land were regularly opened up to white settlers. Establish legal ownership of a plot of land, mark out the boundaries, move on, and repeat. The railroads, just coming into existence when Lincoln was a young man, would accelerate the westward movement and complicate the legal issues. Surveying, and then the law, would raise Lincoln out of poverty, with lawyering for the railroads proving especially lucrative.

The previous inhabitants of the land, the Native Americans, were dismissed as "savages" who had only vague concepts of property and ownership. The very fact that the settlers were able to wrest the land away from the Indians was offered as proof that the settlers were superior and deserving. Some whites decried the violence that so often accompanied the land seizures and were saddened by inhumane treatment of the Indians, but almost without exception whites saw the land transfer as both inevitable and proper. Lincoln, whose grandfather had been killed by an Indian, was no exception.

Lincoln's brief surveying career began in 1833, three years after he had moved from Indiana to Illinois. He was in a deep financial hole, having invested in a failed general store, but recommendation by friends gained him a position as an assistant to the surveyor of Sangamon County, then booming with land speculation. He immediately applied himself to learn the rudiments of the surveying trade, using a couple of standard textbooks he was able to obtain. These books took him substantially past the arithmetic of his youth, to geometry and trigonometry, and to logarithms as an aid to calculation. There are reports of his devoting considerable study to mastering the craft, which might entail such tasks as measuring the area of a plot of land to assess its value, given a price per square foot. This would require establishing the boundary lines, given by compass bearings; subdividing the plot into simple geometric figures such as triangles or parallelograms; measuring the side lengths of these figures, either directly, on the ground, or deductively, by geometric or trigonometric reasoning; performing appropriate multiplications to find the areas of the figures; and then adding these numbers to get the total area.

Lincoln performed at least 14 surveys during 1834-36 and was respected for his work. The income greatly improved his financial situation. At the same time, his interest in politics and the law was growing. He was elected to the Illinois State Legislature in 1834 and admitted to the bar in 1836. He soon ceased surveying.

It was not until more than a decade later, about age forty, that Lincoln made a further step up the ladder of mathematical knowledge. He was again seeking to advance his career, but this time he sought assistance from the more abstract side of mathematics. He had retired from Congress, after serving one term in the House of Representatives. In Congress, courtrooms, and legal documents he had observed men making arguments, and he came to wonder just what it meant to "demonstrate," to convince an audience of the certainty of a proposition. He had gleaned that such certainty was a central feature of mathematics in general, and that the *Elements* of Euclid in particular was considered by many to be the epitome of demonstrative reasoning. Armed with the knowledge of practical geometry he had picked up in his surveying days, Lincoln proceeded to tackle Euclid.

In those days in rural Illinois the need for adjudicating legal disputes outpaced the supply of full-time courts and resident lawyers. This problem was solved by the circuit riding system. A presiding judge and a crew of lawyers would travel about a wide geographical area, resolving litigation at each scheduled town in turn. Lincoln was one of these itinerant lawyers, and when he was reading Euclid, about 1850, he took the *Elements* with him on the circuit, studying whenever he had the opportunity, even, as his partner William Herndon later reported, while his fellow circuit riders were snoring "volubly" in the same room.

Euclid's *Elements* has been compared to the Bible in its wide distribution and pervasive influence, and like the Bible it has been published in a multitude of different forms. The *Elements* as it first appeared in ancient Greek, about 300 BCE, consisted of thirteen "books" (more like a chapter in modern parlance). The first six books cover the basics of plane (i.e., two-dimensional) geometry, and for many decades constituted the essentials of a school geometry course in the United States. The next four books are chiefly concerned with numerical issues, such as prime numbers, while the last three books treat solid geometry. Lincoln claimed, in an 1860 autobiography written for his presidential campaign, to have "studied and nearly mastered the Six-books of Euclid." By his day there were several editions of the *Elements* available in English, giving preeminence to the first six books, but it is not known precisely which edition he used.

Looking, for example, at an English-language *Elements* that Lincoln conceivably could have used, published in New York in 1843, we find it beginning with definitions of basic concepts, such as "a line is a length without breadth" and "an acute angle is that which is less than a right angle." It then proceeds to "postulates," where there is "something required to be done, which is so easy and evident that no person will hesitate to allow it," such as "a straight line may be drawn from any one point to any other point." This is followed by "axioms," each of which is "a self-evident proposition, requiring no formal demonstration to prove the truth of it." An example of an axiom being "things which are equal to the same are equal to one another." Upon this foundation the "theorems" or "demonstrative propositions" are built up, using logical reasoning. Proposition XLVII, much celebrated as the Pythagorean theorem, is a highlight of Book I: "In any right angled triangle, the square which is described upon the side subtending the right angle, is equal to the squares described upon the sides which contain the right angle."

Lincoln's active pursuit of mathematical knowledge seems to have come to an end with his reading of Euclid, but there can be no doubt that, despite his minimal formal schooling, his acquaintance with mathematics was distinctly above average for his time. Some observers have made much larger claims. In their 2010 book *Abraham Lincoln and the Structure of Reason*, David Hirsch and Dan Van Haften, an attorney and electrical engineer, respectively, assert that Lincoln's reading of Euclid had a profound effect on all his subsequent legal arguments and political speeches. Following Proclus, a Greek commentator on Euclid from the fifth century, they see a common rhetorical structure to almost all of Euclid's logical arguments: enunciation, exposition, specification, construction, proof, and conclusion. Lincoln, say these authors, internalized this structure, made it a part of his standard arsenal, and used it to craft the series of compelling demonstrative arguments, both famous and more obscure, that he would make over the rest of his career. Euclid, they conclude, was a major source for Lincoln's rhetorical success. In offering this conclusion they compete with other Lincoln commentators who have proposed alternative explanations for his literary mastery, most notably his familiarity with Shakespeare and the Bible.

Whether or not one is convinced that Euclidean structure permeates Lincoln's later writings and speeches, there can be no doubt of the presence of Euclidean language, most famously in the Gettysburg Address, where he reminded the audience that the nation was "dedicated to the *proposi-*

tion that all men are created equal." This was a truth that had been declared to be "self-evident" in the Declaration of Independence, making it an axiom in the logic of that document. In the political debates of the 1850s Lincoln had repeatedly pointed out how uncomfortable the slave apologists were with the Declaration of Independence, noting that some were reduced to claiming that it was based on "self-evident lies."

In the nineteenth century, two primary justifications were offered for studying mathematics: it is useful in practical affairs, and it trains the mind. Lincoln exemplified both, to an unusually clear degree. But he also exemplified another feature of the place of mathematics in American society: the gulf between the general public and those exploring the research frontier. He, like almost all Americans then and now, had no conception of mathematics as a living, growing activity. In creating his cyphering book, and in his studies of surveying and Euclid, he was learning what appeared to him as a fixed body of incontestable knowledge. For Lincoln the rule of three was a natural law, eternally and universally valid. Euclid all the more so.

But unbeknownst to Lincoln, at the very time that he was looking to Euclid as the epitome of unquestionable certainty, a small network of advanced mathematicians in Europe were indeed questioning it, resulting in the revolutionary invention of what we now call *non*-Euclidean geometry. For about a century scholars had wondered whether it is indeed self-evident that "two straight lines, which intersect one another, cannot be both parallel to the same straight line"—Euclid's fifth postulate—as phrased in an edition that Lincoln could have read. Remarkably, it was found that a logically consistent geometry can be constructed if this axiom is violated. It then became conceivable that Euclid's geometry might not necessarily offer the best description of physical reality, as had been confidently assumed. Physicist Albert Einstein would fully exploit this new standpoint in his general theory of relativity in the twentieth century. In the United States, the development of non-Euclidean geometry remained largely unknown even to the most mathematically well educated until the 1870s, when the founding of Johns Hopkins University brought enlightenment to a small cadre of specialists.

Lincoln was much interested in scientific and technological progress, twice speaking publicly, during 1858–59, on "Discoveries and Inventions." He took a historically expansive view of this topic, lauding advances both ancient and modern: phonetic writing, printing, patent laws, spinning and weaving, the discovery of America, the making of iron tools, the wheel and axle, agriculture, the steam engine. He indulged in tentative prophecy: "quite possibly one of the greatest discoveries hereafter to be made, will be the taming, and harnessing of the wind." But he did not mention mathematics, whose revolutionary developments were invisible to him, as they were to almost all Americans.

4

Textbook Messages

Catherine Beecher and Joseph Ray

Generally speaking there seems to be no very extensive sphere of usefulness for a single woman but that which can be found in the limits of a school-room.

Catherine Beecher

L YMAN BEECHER, a prominent Calvinist clergyman, left Boston with his family in 1832 to become the first president of the Lane Theological Seminary in Cincinnati, Ohio. Within a year, Catherine, his eldest child, founded the Western Female Institute, hired her sister Harriet to aid in running it, and published an arithmetic textbook. At nearly the same time, elsewhere in Cincinnati, instruction was commencing at the new Woodward High School, where one of its teachers, Joseph Ray, would also publish an arithmetic textbook, in 1834. The Western Female Institute soon folded, and Catherine Beecher's arithmetic book sold poorly, but she would go on to become an influential voice on the role of women in American life. Harriet Beecher, after marrying Calvin Stowe, a clergyman colleague of her father, in 1852 published the novel *Uncle Tom's Cabin*, which would eventually sell more copies worldwide than any book except the Bible. The textbook by Joseph Ray became the foundation of a publishing juggernaut long outlasting the author's death in 1855, while the Woodward High School has continued to the present day. It counts among its graduates Eliakim Hastings Moore, the sixth president of the American Mathematical Society, from whom are descended, by the genealogy of doctoral advising, more than 21,000 mathematics PhDs. Another Woodward graduate was William Howard Taft, twenty-seventh president of the United States.

Excepting New Orleans, Cincinnati in the 1830s was the largest city in the country west of the Appalachian Mountains, with a population of about 25,000 at the beginning of that decade. Like Salem a few decades earlier, Cincinnati was having a brief moment of national prominence. It was considered, by most Americans of European heritage, to be at the edge of civilization, but its location at the confluence of the Licking and Ohio Rivers made it both accessible to migration from the east and convenient as a jumping-off point for venturing farther west or south.

Ohio had been carved out of the Old Northwest, where slavery had been outlawed by the Northwest Ordinance of 1787. But Cincinnati lay just across the Ohio River from Kentucky, where slavery was legal, so that issues connected with that institution were frequently forced upon the attention of city residents. For the free white population there were abundant opportunities for making money, especially in land speculation and hog raising, the latter giving rise to the nickname "porkopolis." William Woodward's fortune, which built the school named after him, was largely derived from a land seizure that had inspired aggrieved Indians to "massacre" a white family in 1801, two years before Ohio became a state. Woodward had become guardian of the surviving female child, and when the orphaned girl came of age, he married her, securing ownership of her late father's property.

The growing contingent of wealthy Cincinnati residents had increased demand for schools in the 1820s. At first this produced only private schools to serve their own children, but in time the general educational agitation would result in the Cincinnati public school system, the first such system in the state. The Woodward school originated, like Catherine Beecher's Western Female Institute, as a private institution. Mr. Woodward had originally envisioned supporting a grammar school, specifically for children of the poor, but influential residents were eager for something of a more advanced nature, inducing the founder to alter the terms of his bequest to make it a high school. Just a few years after Woodward's death, in 1836, it became Woodward College and High School, awarding both bachelor's and master's degrees along with high school diplomas. In 1851, as distinctions between colleges and high schools were solidifying nationally, the decision

Joseph Ray. Image AL04174, courtesy of the Ohio History Connection

was made to again make it strictly a high school, and to merge it into the city's public school system. Joseph Ray was named the principal.

Ray had grown up on a farm in the slim panhandle of land separating Pennsylvania and Ohio. This was a portion of Virginia that had much less allegiance to slavery than the rest of the state, contributing to its splitting off, during the Civil War, to form part of the new state of West Virginia. Like his near-contemporary, Abraham Lincoln, Ray could attend school only during the seasons when his manual labor was not needed, but Ray's region of the country was somewhat better supplied with school-teachers than Lincoln's. Ray even attended an institution calling itself an "academy," just over the state line in Pennsylvania. At age fifteen he was studying algebra, geometry, and surveying, and one year after that he was working as a teacher himself, near the family farm. A doctor with whom Ray apprenticed arranged for him to attend the Medical College of Ohio, in consequence of which he became Joseph Ray, MD, and was often referred to as Dr. Ray. He practiced as a physician only briefly before taking a position at the newly opened Woodward High School in 1831. He remained there the rest of his life, as a teacher of mathematics and as an administrator.

Catherine Beecher, as with most women of the time, had less formal education than Joseph Ray, but the stimulating social environment in which she had grown up in the Northeast had prepared her to conduct a lifelong intellectual career. Her father, Lyman, a graduate of Yale College, had led churches in New York and Connecticut before his stint in Boston. He was a prominent and effective voice among those debating the future of Protestant Christianity in early nineteenth-century America, a participant in what historians refer to as the Second Great Awakening. Strict Calvinist doctrine, as promulgated by preachers such as Jonathan Edwards in the First Great Awakening of the mid-1700s, had held that individual salvation was at the whim of an angry God. Beecher was among those evangelical preachers who sought to allow some leeway for personal goodness to affect a person's fate. He hoped thus to counter the attractions of alternative faiths, especially in Boston, where the altogether too broadminded Unitarians held sway. Several of his brood of bright and inquisitive children (12 survived to adulthood) pushed further into theological liberalism but never entirely escaped the imprint of their father.

Catherine at age twenty-one had become engaged to Alexander Metcalf Fisher, a professor of mathematics and natural philosophy at Yale with a growing reputation. He was a leading light of the as yet tiny community of American academic mathematicians, the community that Nathaniel Bowditch hovered around but declined to join. When Fisher tragically died in a shipwreck on the coast of Ireland in 1822, on his way to observe European institutions of higher learning, Fisher's family invited Catherine to tutor his younger siblings and gave her access to his private papers. In reading his diaries and his unpublished scientific and mathematical jottings dating back to his boyhood, she found within herself a greater attraction to the life of the mind, and more intellectual power, than she had previously realized. Preparing her lessons for the Fisher children, she ended up reading and understanding books on arithmetic, algebra, geometry, and logic.

This gave her the idea that teaching might be a profession at which she could succeed.

Before coming to Cincinnati, Catherine had founded the Hartford Female Seminary in Connecticut. The school was highly successful, for a time. As she supervised its operation, she discovered further weaknesses in her own education. This led her to closely examine books used in other schools, and being the critical thinker that she was, she found imperfections. In two cases she decided to write her own books, which she could then use in her school. Her book on mental and moral philosophy was privately printed in 1831, and her arithmetic textbook, probably the first mathematics text written by a woman in the United States, was first issued privately in 1828 and then published, with many improvements, in 1832. But unlike mental and moral philosophy, where she was writing of matters that touched her profoundly, arithmetic was a topic she found "both the most difficult and most uninteresting." She only gave it attention because of "its practical usefulness in my profession." In short, arithmetic was expected to be taught in any institution that called itself a school, and so she felt compelled to deal with it.

Catherine used the popularity of her school to establish herself in Hartford society, and she began as well to envision a more ambitious program for the education of women generally. Having largely given up the idea of marrying, she increasingly saw teaching as an alternative to marriage, not only for herself, but for a whole national contingent of middle- and upper-class women, who would inculcate their pupils with strong moral principles. It was much in the spirit of her father's evangelical preaching, without being under the official sponsorship of a church. Schoolteaching was not yet the female-dominated profession that it would become; Catherine Beecher was one who helped make it such.

By 1830 both Catherine and her father were reaching professional dead ends in New England. Hartford families were annoyed at Catherine's attempt to combine instruction in genteel behavior with rigorous moral strictures, while Lyman was seeing little payoff from his evangelistic struggle with the Unitarians in Boston, where he had moved in 1826. Both father and daughter found the prospects of a new beginning in Cincinnati alluring. It was, opined Lyman, "the London of the West." At first Catherine advanced rapidly in the local hierarchy. The Western Female Institute got off to a strong start, and she established a friendship with William McGuffey, a prominent local educator.

Catherine's *Arithmetic Simplified* was published in Hartford just as she moved to Cincinnati. Her innovations in this book were not drastic. It covered all the standard topics found in other books of the time and in Abraham Lincoln's cyphering book: adding, subtracting, multiplying and dividing whole numbers and fractions; proportion (the rule of three); simple and compound interest. The most advanced topics were extraction of square and cube roots and arithmetic and geometric progressions. The book also included problems using both federal and English money. She claimed to have organized the material to make it useful to a wide variety of ages and preparation levels, and to have provided fuller explanations, compared to other books, of the rules offered for solving problems. She dismissed one currently popular arithmetic book as rife with "mystical performances" for its failure to provide such explanations. She also asserted, plausibly, that her approach had benefited from repeated experimentation in her school.

But *Arithmetic Simplified* was not a success. Catherine blamed the publisher for being unwilling to fully promote a book written by a woman. She revised the text slightly, in 1835, as the *Lyceum Arithmetic*, using a Boston publisher and replacing her name on the title page with "an experienced teacher." Nevertheless, this book too failed to be profitable.

Beecher would write other books, but never again in mathematics. Her most widely influential work was *Treatise on Domestic Economy*, originally published in 1841 and reprinted many times. This was her ambitious attempt to put a woman's work in the home on the same footing as academic subjects: "Why are not the application of these laws to the management of infants and young children as important to a woman as the application of the rules of arithmetic to the extraction of the cube root." In this book, and in other writings, Catherine's aim was not to argue for equality of the sexes, discussion of which she considered "frivolous and useless." Rather, she supported the view that men and women

Catherine Beecher at age fifty-seven. Image WHi-3127, Wisconsin State Historical Society

were responsible for different aspects of life. Thus she was unenthusiastic about giving women the right to vote, but she strongly urged that women's work, inside and outside the home, be properly recognized for its importance, and that women who chose not to marry be educated for economic independence. Although in running her schools she took full advantage of, and even extolled, the fact that women teachers were willing to work for less pay than men, toward the end of her life she supported equal compensation.

Life in Cincinnati for Catherine and the other Beechers did not remain smooth for long. The city had attracted many German Catholic immigrants who were not pleased by Lyman's strident attacks on their faith, an activity he considered required of a devout Protestant. Further tension was produced by Catherine's unhidden belief that the city's New Englanders were a cut above those who had come from other regions. And looming over all social relations in the city was slavery. While Joseph Ray seems never to have publicly commented on the subject, Lyman Beecher and his children were not so reticent. They sought to stake out an "intermediate" position, decrying both the institution of slavery and those who called for immediate emancipation. But as Abraham Lincoln was later to discover, such moderation was no haven, and events in Cincinnati made neutrality increasingly difficult.

In 1834 a group of Lane Seminary students, disgusted by Lyman Beecher's timidity, dramatically departed Cincinnati for northeast Ohio to found Oberlin College, where they could promote their radical abolitionist agenda unimpeded. Then in 1836 proslavery mobs destroyed the presses of a Cincinnati newspaper with abolitionist sympathies, with many prominent citizens tacitly approving. Harriet in particular was deeply shocked, while still professing to oppose "the excesses of the abolition party." The following year Catherine published a book in which she declared that Northern abolitionists were wrong to support the socially disruptive policy of immediate emancipation, and counterproductive to hurl invective from their comfortable abodes at Southerners who would feel the brunt of that disruption. She and the other Beechers at that time had difficulty conceiving of a racially integrated future for the United States, instead seeing "recolonization" to Africa as a practical solution. She and her father retained this stance into the 1840s, even as several of her brothers, now becoming prominent ministers in their own right, more fully embraced abolition. Recolonization continued to be mentioned favorably in Harriet's *Uncle Tom's Cabin*, written after the Fugitive Slave Law of 1850, by which time the Beechers had all left Cincinnati for points north. Harriet wrote much of the novel in Brunswick, Maine, while her six children were looked after by Catherine.

Meanwhile, Joseph Ray was demonstrating how to navigate both Cincinnati society and textbook publishing. Although Catherine Beecher continued to rely on East Coast publishers for her books, mathematical and otherwise, Cincinnati was rapidly becoming a major publishing center. Joseph Ray, who quickly became well accepted in the city for his educational acumen and social polish, was well positioned to take advantage. In the early 1830s the small publishing firm of Truman and Smith had begun seeking local authors to write school textbooks. With the now-familiar hierarchy of graded schools just coming into being in the United States, they had the crucial idea of creating not mere stand-alone books, but series of books, each title building on its predecessor to carry a student through an entire school career. The business advantages of this scheme, ignored by Beecher, proved substantial. The advent of such inexpensive textbook series would mark the beginning of the end of the cyphering book tradition that had nurtured earlier students, such as Abraham Lincoln.

One of the first prospective authors that Truman and Smith approached was Catherine Beecher, who was asked to edit a series of reading textbooks. But Beecher, wishing to focus on higher levels of the school curriculum, turned down the offer and recommended her friend William McGuffey, then a professor of mental philosophy at Miami University, 35 miles from Cincinnati. McGuffey, who would briefly be a colleague of Joseph Ray at Woodward College in the 1840s, proceeded to produce his celebrated series of *Eclectic Readers*. The early volumes featured simply worded vignettes constructed by McGuffey himself, while later volumes contained selected passages from writers he considered to be leading lights of English literature, among whom he loyally ranked both Lyman and Catherine Beecher. All the selections served to promote moral virtue and patriotism in a manner immensely appealing to white Protestant America. Truman and Smith, and successor publishing companies, proved adept at promoting these books, which would sell seven million copies by 1850, and many millions more over the rest of the nineteenth century.

Even before McGuffey produced his first reader, in 1836, the Truman and Smith Company was launching a corresponding series in arithmetic, with Joseph Ray as the author. His *Little Arithmetic* of 1834 was only 67 pages and covered far less of the subject than Catherine Beecher's text, but the series plan was already evident on the title page: "Introduction to Ray's Eclectic Arithmetic." Within a few years *Little Arithmetic* was featured as the middle volume of "Ray's Arithmetical Course in Three Parts," beginning with *Ray's Picture Arithmetic*, even more basic than *Little Arithmetic*, and closing with *Ray's Eclectic Arithmetic*, nearly identical in content to Beecher's book.

Like the McGuffey series, the Ray series proved enormously popular, outselling all other arithmetic books of the time. New editions of Ray's books were issued frequently, with no particular concern for precise numbering. By 1850 some title pages were featuring the striking but absurd claim of being the "1000th edition." The profitability of the Ray and Mc-Guffey books would be central to a series of mergers and buyouts in the publishing industry, which would lead to the foundation in 1890 of the American Book Company of New York City, a behemoth of the textbook trade.

In 1849, Book 1 of Ray's algebra series appeared, soon followed by Book 2. By the time of Ray's premature death from tuberculosis in 1855 his name had become so synonymous with mathematics textbooks that the wise business decision was to continue to reprint the books he had written and attach his name to additional books written by other hands. Thus "Ray's Mathematical Series" would eventually include books on geometry, trigonometry, surveying, astronomy, and calculus, all written after his death.

Many books would continue to be published under Joseph Ray's name into the twentieth century. Indeed, Ray's arithmetic books have received a new lease on life in the twenty-first century. Both McGuffey's readers and Ray's arithmetic books have found favor with some homeschooling enthusiasts for their nineteenth-century aura of mental rigor and moral rectitude. A few of Ray's many word problems appeal to fundamentalist Christians: "From the creation of the world to the flood was 1656 years, thence to the siege of Troy 1164 years, thence to the building of Solomon's Temple 180 years, thence to the birth of Christ 1004 years. In what year of the world was the Christian era?" But most of his problems, like Beecher's, are comfortably bland, although perhaps inspiring nostalgia for simpler bygone days:

A steamboat that can run 15 mi. per hr. with the current, and 10 mi. per hr. against it, requires 25 hr. to go from Cincinnati to Louisville, and return; what is the distance between these cities?

A person bought a number of sheep for $80; if he had bought 4 more for the same money, he would have paid $1 less for each; how many did he buy?

Until just a few years before her death in 1878, Catherine Beecher would continue to travel, lecture, and write on religion, female education, and domestic science. Mathematics was never a deep interest but merely a small part of her broad social vision for education. Her arithmetic books never offered a clear rationale as to why a student should learn the material therein, except that arithmetic was an accepted part of a school education. Joseph Ray, in contrast, was undistracted by wider social issues, and explicitly presented arithmetic as a tool of business and algebra as a means to discipline the intellect. Beginning as a generalist, undecided between teaching and medicine, he had ultimately become a mathematics education specialist. This would prove to be the wave of the future.

5

Learning to Count

J. Willard Gibbs

1841

The characteristic of Gibbs' work was the amount he could get out of very general assumptions, few in number. . . . Of course, his phase rule is done in half a page just counting constants. That's all it amounts to, really.

E. B. Wilson

I N MARCH 1841 THE SUPREME COURT of the United States, after hearing an impassioned argument by former president John Quincy Adams, affirmed a lower court ruling that the African passengers of the Spanish ship *Amistad* were entitled to freedom. This decision relied crucially on the success of Professor Josiah Willard Gibbs of Yale College in establishing communication with the Africans. They were thus able to explain that they had not been long resident in bondage in Spanish-ruled Cuba, as claimed by the ship's captain, but in fact had been recently transported across the Atlantic after capture in Sierra Leone, a state of affairs in violation of explicit treaty obligations of Spain. Gibbs had managed his linguistic feat by first learning from the Africans how to count to 10 in their language, and then, by repeatedly speaking this sequence as he strolled among seamen along the docks of New York Harbor, finding a young man who could speak both the African language (Mende) and English. At the time of the Supreme Court decision, Gibbs was the father of a two-year-old child, also named Josiah Willard Gibbs, who would grow up to grace the Yale faculty as a professor of mathematical physics. The younger Gibbs is often considered the greatest American scientist of the nineteenth century and indeed one of the world's greatest scientists. Unlike the case of Nathaniel Bowditch, the European admiration for this man exhibited no trace of "not bad, for an American."

≡≡≡≡

Josiah Willard Gibbs the elder, known to family and friends as Josiah, was descended from a highly educated lineage with roots in the earliest days of colonial North America. Within his quiet life of scholarship, largely devoted to studying languages ancient and modern, the public nature of the *Amistad* incident was an outlier. His role was noted in early accounts of the case and received considerable attention in the 1942 biography of his son by poet Muriel Rukeyser, but the senior Gibbs has never achieved fame. An opportunity for greater recognition might have occurred in 1997, with the release of Steven Spielberg's *Amistad*, but the Gibbs of this movie is a comic foil, with his linguistic knowledge depicted as partly fraudulent and some of his contributions to the *Amistad* case attributed to others. Still more recently, Josiah Gibbs has been lauded at Yale as an authentic voice of nineteenth-century abolitionism, but his name has been rather lost in the hubbub over Yale's past connections with slavery, notably the controversy over removing the name of staunch slavery proponent John C. Calhoun (class of 1804) from one of its constituent colleges.

Josiah Willard Gibbs the younger, usually known as Willard, followed his father in both the quiet dignity of his life and in his rigorous scholarship, but without any instance where his name was brought before a large public. His fame is deep and abiding among mathematicians, physicists, and chemists attuned to history. A handful of literary intellectuals, from his time to ours, have also found Gibbs fascinating. But the general educated public has remained largely unaware. The public knows Einstein, and Einstein spoke glowingly of Gibbs, but the public does not know Gibbs.

Yale, in the first half of the nineteenth century, was the largest and most influential college in the country, its graduates far more inclined than those of the more insular Harvard to attempt to transplant their institution's spirit to newly founded colleges in the west. That spirit was for many years notably conservative, as embodied in the Yale Report of 1828, written in response to complaints that educational progress was being stifled by the classical curriculum of Greek, Latin, and mathematics. The Yale authorities championed that curriculum, with perhaps a few flourishes allowed. The overriding aim was to furnish the minds of the elite with the powers needed to perform their supervisory roles in society. It was not originally part of the college's official mandate to open new realms of knowledge, for instance, by analyzing the grammar of obscure African languages or by attempting to deduce the observed laws of thermodynamics from more basic principles.

The elder Gibbs was born in 1790 in Salem, Massachusetts, when that town's fortunes were still on the rise. Coming from a family where college was already a tradition for the men, he received a superior school education and proceeded to Yale. He excelled in all parts of that college's classical curriculum, including mathematics, at a time when there was no such thing as a major; all students took the same array of courses. After graduating in 1809 he became fully licensed to preach by attending a seminary in Andover, Massachusetts, but his encounter there with rigorous German biblical exegesis, and his fascination with languages, inclined him to be a theological scholar rather than a minister. His son too would be greatly impressed by German learning. The father eventually became professor of sacred literature at his alma mater. He published a Hebrew lexicon and numerous modest notes on comparative grammar. In 1857, four years before he died, he organized his lifetime of linguistic cogitation into *Philological Studies with English Illustrations*. Many of the principles he espoused, he diffidently declared, "are now current in our schools of learning," but he hoped nevertheless "that the publication may not be amiss." He admired, according to one observer, "genial and tasteful appreciation of shades of thought." But for all his hesitance to make dogmatic pronouncements, he had some strong opinions. His assistance to the *Amistad* captives was unusual for being public, but entirely in character. The same observer noted that "His heart was deeply affected by the wrongs inflicted in this country upon the African race."

Josiah Willard Gibbs, Yale College, 1858. Images of Yale individuals, ca. 1750–2001 (inclusive). Manuscripts and Archives, Yale University

By the early 1850s, as the younger Gibbs was growing up, Yale had unbent enough to establish what became known as the Sheffield Scientific School, admitting that science was deserving of greater recognition as an academic pursuit. The scientific students, however, were for many years considered inferior to those pursuing the traditional Yale degree. Willard Gibbs followed in his father's footsteps and earned that traditional degree in 1858, excelling in both languages and mathematics.

Another Yale innovation of the 1850s was congenial to Willard's temperament and inclinations: the offering of graduate study and the awarding of advanced degrees. In 1861 Yale became the first American institution to award a PhD. As the country descended into the ghastly conflict of the Civil War, failing utterly to solve the problems of racial injustice that had so troubled the elder Gibbs, young Gibbs, rather than following many Yale classmates into military service, "withdrew" (as poet Rukeyser would later describe it) into the academic world he would inhabit for the rest of his life. He studied chemistry, physics, and astronomy, and he wrote a thesis on an engineering topic, for which he was awarded a PhD in 1863: "On the Form of Teeth of Wheels in Spur Gearing."

Because of his father's political views, Willard Gibbs's avoidance of the Civil War, and more generally his apparent failure to express any opinions or take any stands on social questions, has given pause to later commentators. His two full biographers, Muriel Rukeyser and Lynde Wheeler, have explained his behavior in the 1860s on practical grounds. Willard's health had never been considered robust, with tuberculosis sometimes feared. And with his father's death in 1861 he became the main support of his two unmarried sisters. Later in life he became totally engulfed in his work. In view of the magnificence of that work, all is forgiven. Others have been more troubled, worrying that Gibbs exemplifies the pure scientist sailing blissfully and irresponsibly above the struggle, ignoring the realities of life. It has been noted that other great scientists, Einstein for one, have not let their scientific work prevent them from speaking out on social issues.

With his degree in hand, Gibbs took an appointment as a tutor at Yale, called on to teach all the subjects for undergraduates in their first two years. At first he was primarily assigned to teach Latin. Student life was evolving in those days, as young men with definite professional aims to become lawyers, doctors, or ministers were increasingly replaced by those attending purely for social prestige. There are vague reports that Gibbs was not comfortable dealing with such intellectually disinclined customers. Later in his career he would confine his teaching responsibilities to graduate students and mathematically advanced undergraduates.

After his three-year tutorial appointment had expired, Willard set off for a tour of European universities. He was able to make this trip, taking his two sisters, owing to the comfortable financial status of the Gibbs family, especially buoyed by his father's prudent investment in railroad stocks, an industry then undergoing a vast expansion, despite the chaos surrounding the Civil War. He spent time at universities in Berlin, Heidelberg, and Paris, over a period of three years, attending lectures by and reading deeply in some of the great masters in science and mathematics of that era: Bunsen, Chasles, Helmholtz, Kirchoff, Kronecker, Kummer, Liouville, Weierstrass. He learned who was who in the world of European science, a knowledge that would serve him well in future years.

The research university as we know it today, in which scholars train succeeding generations in the skills needed to advance knowledge in a specialized field, had been pioneered in France during the Napoleonic era and then elaborated by German universities in subsequent decades. The United States was far behind in this development, but some American institutions, most notably Johns Hopkins University, would adopt the German model during Gibbs's lifetime. Conservative Yale, despite awarding the occasional PhD, long remained committed to an older vision of higher education, and Gibbs himself, although he devoted the greater part of his life to research, never sought to remake Yale. Research for Gibbs, as it had been for his father, was a personal calling, not the professional team-building phenomenon it would become in the twentieth century. He never aggressively sought out students to assist in or to further his research projects, a practice that would become a hallmark of the research university. He just took on whoever showed up.

But merely by example, Gibbs did inspire a small group of remarkable students. In 1902, shortly before his death, he observed that in over 30 years of teaching, he had had only six students who were adequately prepared to understand his lectures in mathematical physics. Two he named specifically were Edwin Bidwell Wilson, to whom this remark was offered (see chap. 13), and a graduate of Cincinnati's Woodward High School, Eliakim Hastings Moore, who by 1902 was the chairman of the University of Chicago's mathematics department and president of the American Mathematical Society (see chap. 10). The identities of the other four may be guessed at. Likely they included Lynde Wheeler, who would become Gibbs's biographer. Other plausible candidates are Yale physics professor Henry Bumstead, radio pioneer Lee de Forest, and Irving Fisher, a noted economist.

After his return from Europe in 1869, Gibbs never again left the United States and rarely left New Haven. He lived the rest of his life in the house in which he grew up, along with his two sisters, one a spinster and the other married to the Yale College librarian. Yale appointed Gibbs to a position as a professor of mathematical physics, with no salary. The college only deigned to compensate him after Johns Hopkins threatened to hire him away about 1880, although even then Gibbs accepted less than Hopkins had been offering. His independent wealth allowed him to make such a decision, and since Yale gave him no responsibility for introductory teaching, he had leisure to probe deeply into any aspects of science he found most appealing. From the beginning of his career, Gibbs displayed an audacious willingness to rethink the foundations of whatever topic he was investigating, while at the same time being careful not to claim more than he could securely establish. His publications were few but polished to a condition of intimidating concision.

Thermodynamics was a topic that Gibbs began to investigate in the early 1870s. Provoked substantially by the growing use of steam engines in industry, the understanding of heat among European physical scientists underwent a revolution in the nineteenth century. In the late eighteenth century, the prevailing notion had been that heat was a fluid, called caloric. A hot substance had more of it, a cold substance less. When determined experimental scrutiny failed to support this concept, however, an alternative view gradually came to the forefront: the mechanical theory of heat. A hot substance was one in which the constituent particles were vibrating more rapidly than in a cold substance. By the 1860s this theory had received considerable support and had been codified into two fundamental laws.

The first law of thermodynamics asserted that in an isolated physical system, energy is conserved, no matter what contortions the constituents of the system may undergo. Mechanical energy can be transformed into heat energy (as in rubbing two sticks together), and heat energy can be transformed into mechanical energy (as in a steam engine), but if careful accounts are kept, one finds that no energy is created or destroyed. The second law, subtler but well confirmed by experience, was that any transformation of energy in an isolated system reduces the availability of energy for subsequent transformations. Only shortly before Gibbs came on the scene, this second principle had been tentatively proposed by the German physicist Rudolph Clausius. Clausius declared that when heat flows from hot bodies to cold bodies, a quantity he called entropy increases. Because observation overwhelmingly supports the view that heat will naturally flow from a hot body to a cold body, but not the reverse, Clausius formulated the second law of thermodynamics as the claim that the entropy of an isolated system always tends to increase. When ice is placed in a glass of water, the ice always melts, while the water never freezes. Thermodynamics was seen to be fundamentally concerned with irreversible processes.

Gibbs came to intellectual maturity as the world's scientists were attempting to assimilate these ideas. One of his major contributions was to bring a geometric point of view, showing how making graphical depictions of the variables related to thermodynamics (pressure, volume, temperature, energy, entropy) improved understanding. His geometric reasoning is exemplified by the title of one of his papers: "A Method of Geometrical Representation of the Thermodynamic Properties of Substances by Means of Surfaces." If the values of two quantities are depicted in a two-dimensional plane, then the variation of a third quantity in relation to the first two can be depicted as a surface of hills and valleys lying above that plane. Gibbs used pictorial representations to offer a more comprehensive understanding of phase transitions among solid, liquid, and gaseous forms of a substance. He applied his improved grasp of thermodynamics to derive important consequences for other physical phenomena, especially chemical and electrochemical. His 1878 paper "On the Equilibrium of Heterogeneous Substances" has been much celebrated as providing the foundation of the field of physical chemistry. His "phase rule" proved especially useful, providing a simple numerical summary of the interrelationship among variations of temperature, pressure, and phase transitions within a chemical mixture. Specifically, the phase rule asserts that for a chemical mixture the number of thermodynamic variables (such as temperature and pressure) that can be varied independently is given by the formula $C + 2 - P$, where C is the number of distinct chemical components (say, oil and water) and P is the number of phases of matter (gas, liquid, solid) able to be realized within the particular problem context. But some chemists have complained that Gibbs's approach was less fruitful than it might have been, owing to its high level of abstraction and Gibbs's lack of direct knowledge of experimental chemistry.

Gibbs published his thermodynamic results in the *Transactions of the Connecticut Academy of Arts and Sciences*, a local New Haven scholarly publication, often deemed by later commentators as "obscure." One might presume that he chose this publishing venue to minimize the effort of broadcasting his ideas, but this impression is belied by the further fact that Gibbs carefully mailed copies of his papers to a substantial and carefully constructed roster of the most prominent European and American scientists, demonstrating his clear knowledge of those most likely to understand his work. In this way, he gained enthusiastic support from the great Scottish mathematical physicist James Clerk Maxwell, who alerted colleagues in Great Britain to the importance of Gibbs's work.

After his foray into thermodynamics, Gibbs turned his attention to Maxwell's magnificent unification of the multitude of phenomena associated with electricity and magnetism. As with thermodynamics, theoretical developments in this field were substantially motivated by ongoing practical applications, in this case telegraphy. Maxwell had laid out his ideas in full detail in his *Treatise of Electricity and Magnetism* of 1873, before dying of cancer in 1879 at the age of forty-eight. In the course of developing his theory, Maxwell had had recourse to a mathematical structure known as quaternions, initiated in the 1840s by the Irish mathematician and physicist William Rowan Hamilton.

Hamilton had discovered that it was possible to create an algebraic system of pairs of numbers (a, b) in such a way that these pairs could be added, subtracted, multiplied, and divided just like ordinary one-dimensional numbers. In fact, an isomorphic copy (as mathematicians would now say) of the ordinary numbers was contained within Hamilton's pairs, namely, all the pairs of the form $(a, 0)$, with the second component always 0. But Hamilton avoided what might seem the obvious scheme for multiplication, $(a, b) \times (c, d) = (ac, bd)$, in favor of the more convoluted $(a, b) \times (c, d) = (ac - bd, ad + bc)$. This had the notable consequence that $(0, 1) \times (0, 1) = (-1, 0)$. In other words, -1 was endowed with a square root. Earlier generations of mathematicians had been so leery of square roots of negative numbers that they had dubbed them "imaginary." Hamilton's scheme made them seem more concrete. He then set about trying to make a similar algebraic structure out of triples: (a, b, c). Such a structure would have obvious useful applications in physical science.

But Hamilton failed to find an algebra of triples with all the properties he wanted; later mathematicians would demonstrate that no such algebra is possible. But he did find an algebra of quadruples, (a, b, c, d), which he called quaternions. Even this system had one property deviating from ordinary numbers, but he was willing to live with it: the quaternions failed to be commutative; $A \times B$ was not necessarily the same as $B \times A$. By a bit of juggling, the four-dimensional quaternions could be made to serve some of the needs of physicists describing phenomena in three-dimensional space. Maxwell used quaternions in this way in his *Treatise*.

Hamilton was deceased by the time of Maxwell's *Treatise*, but some of his disciples were promoting quaternions with great fervor as the wave of the future in physics and mathematics. Gibbs read Maxwell closely, and he took note of what many other readers seemed to ignore, that Maxwell was not a full-fledged endorser of quaternions. Rather, Maxwell opined that the greatest value of Hamilton's work was the distinction he had made between a vector, an object that could be thought of as an arrow with a length and a direction, and a scalar, a pure dimensionless quantity. From this hint, Gibbs proceeded to develop what he called vector analysis, which proved to be the ideal formalism for expressing physical theories, as well as a fruitful source for pure mathematical generalization and elaboration. Physicists eventually discarded quaternions, seeing them as a mere scaffolding for developing the theory of vectors, although they have retained a place in mathematics for their special algebraic properties.

Gibbs worked up his algebra of vectors in a privately printed pamphlet that he used in teaching electromagnetic theory at Yale in the 1880s. Here again, he distributed this work to leading scientists and mathematicians. In doing so, he incurred the wrath of quaternionic true believers but found favor with those on the winning side. About 1900, he encouraged one of his best graduate students, Edwin Bidwell Wilson, who had recently come to Yale with a Harvard bachelor's degree, to write up the material in more finished form. The resulting book was published in 1901, under Wilson's name, as *Vector Analysis, A Text-Book for the Use of Students of Mathematics and Physics, Founded upon the Lectures of J. Willard Gibbs.* It was a main-stay of college physics and mathematics classrooms across the United States into the 1940s.

Gibbs did not have time to write up the vector analysis book himself because he was preoccupied with his last great project—to systemize the field of statistical mechanics. The basic problem was to try to explain the large-scale behavior of matter (encapsulated in measurements such as temperature, pressure, and volume) as a consequence of the microscale behavior of the tiny particles making up the matter. There was no hope of knowing precisely what all the individual particles were doing, but if one could estimate the likelihood that a certain proportion of the particles were moving in certain ways, then one might be able to somehow average the whole conglomeration of motions. The problem became, in effect, how to appropriately group and count collections of particles. Maxwell had made progress on the problem, the Austrian Ludwig Boltzmann had advanced further, and finally Willard Gibbs, in 1902, just a year before his death, placed his masterful stamp upon the field with his *Elementary Principles in Statistical Mechanics.*

This was, and is, a greatly admired book. Einstein declared it a master-piece, although he also said it was hard to read and admitted that he had not actually read it until after publishing his own work in the field, which duplicated some of what Gibbs had done. In a faithful English translation of *La Valeur de la Science,* by the brilliant French mathematician and phys-icist Henri Poincaré, the Gibbs book is described as "too little read because it is a little difficult to read." Gibbs's student E. B. Wilson gave a looser, punchier translation of Poincaré, which he passed on to his own students: "a little book, little read, because it is a little hard."

The image of the uncompromisingly difficult Gibbs, admired by ex-perts worldwide, has intrigued a select group of American writers. An early instance appears in *The Education of Henry Adams,* privately distrib-uted in 1907 but not published until after Adams death in 1918. In this au-tobiography, the grandson of John Quincy Adams cited Gibbs as "the greatest of Americans, judging by his rank in science." Adams gave a fuller discussion of Gibbs in an essay, "The Rule of Phase Applied to History," completed in 1909 but also not published until 1918. This piece demonstrates Adams's fascination with the metaphorical suggestiveness of scientific ideas, with little appreciation for or interest in the productive interplay between precise logical reasoning and shrewd approximation that characterizes modern physical science. Although Henry's brother, Brooks Adams, reported that Henry discussed his musings on the phase rule with Gibbs's student H. A. Bumstead, the result is a parade of undigested scientific terminology in the service of Adams's increasingly gloomy view of the human condition. For Adams, words such as entropy, critical point, phase, and equilibrium never achieve more than amorphous content. Mathematics was for him a mystical field of endeavor: "The true mathematician drew breath only in the hyper-space of Thought; he could exist only by assuming that all phases of material motion merged in the last conceivable phase of immaterial motion—pure mathematical thought." This entirely misses the decidedly utilitarian spirit of Gibbs approach to mathematics, so essential to his achievement. The importance of Gibbs's rule of phase for science lies not in vague implications but in its explicit numerical character: if certain simplifying assumptions are made, certain precise results follow, and these results can be used to predict specific useful phenomena in the world.

Muriel Rukeyser first wrote about Gibbs in a poem published in 1939, referring to him as a "very great and partly unacknowledged" American cultural figure. Rukeyser's opening stanza draws on an anecdote about Gibbs at a Yale faculty meeting, during which the relative merits of studying languages and mathematics were discussed. Gibbs was reported to have risen up to declare, "Mathematics *is* a language." The rest of the poem returns repeatedly to the words "withdrew" or "withdraw," words Rukeyser saw as defining Gibbs's life.

Following her effort to capture Gibbs in about 100 lines of verse, Rukeyser proceeded to write a full-length biography, published in 1942. The book spends much time on the *Amistad* incident and on Henry Adams; attempts valiantly to draw connections with Abraham Lincoln, Herman Melville, and Walt Whitman; and in general seeks to situate Gibbs at the center of nineteenth-century American life. She explicitly acknowledges her "presumption," as a poet, in endeavoring to describe Gibbs's scientific contributions. But finding that no one else seems willing to take on the task, she persists, seeking thereby to bring about a reconciliation between art and science. Some of those who had been close to Gibbs found the Rukeyser biography distasteful, objecting to what they considered her undisciplined prose, scientific naiveté, and exaggeration of Gibbs's lack of recognition. E. B. Wilson seems to have been unhappy that a young woman of Jewish heritage and leftwing politics would dare to write about a man of conservative temperament from deep-rooted Yankee stock. Eventually one of Gibbs's other students, Lynde Wheeler, countered the Rukeyser biography with *Josiah Willard Gibbs: The History of a Great Mind* in 1951.

A more recent writer, Thomas Pynchon, with greater technical knowledge than either Adams or Rukeyser, has long delighted in seeding his fictional writings with scientific ideas and people. Thermodynamics has been a special interest, from his short story "Entropy" of 1960, written while he was still in college, to his massive phantasmagoric novel of 2006, *Against the Day*, in which one of the main characters studies at Yale, encountering "the kindness and genius of Willard Gibbs." There are numerous references to quaternions and vectors and the "Quaternion Wars" of the 1890s. For understanding the second law of thermodynamics, one might do worse than to read Pynchon:

> "If anything's an irreversible process, cooking is!" lectured Thermodynamics Officer Chick Counterfly, meaning to be helpful, though unavoidably in some agitation. "You can't de-roast a turkey, or unmix a failed sauce—time is intrinsic in every recipe, and one shrugs it off at one's peril."

Lynde Wheeler in his biography offers well-informed descriptions of Gibbs's scientific work, though it is questionable whether he succeeds any better than Rukeyser in enlightening nonscientific readers. What most distinguishes Wheeler's biography from earlier surveys of Gibbs's life is his portrayal of Gibbs as "pre-eminently a happy man." He was happy in his life in New Haven, as circumscribed as it may have appeared to outsiders; he was happy in his Yale colleagues; and, above all, he was happy in his intellectual work. Rather than the man who "withdrew," as described by Rukeyser, Wheeler portrays Gibbs as essentially an intellectual pleasure seeker. This accords with an anecdote recounted by E. B. Wilson about one of his last meetings with Gibbs. Proposing a course for Wilson to teach, Gibbs ended by saying, "I don't want you to do it unless you want to, because everyone should do what he wants to do. He does that better than things he doesn't want to do."

6

Naval Reserve

Charles H. Davis

1857

The greatest single testimonial to the fact-gathering power of science, resting, as it does, on centuries of labor and ingenuity, is probably the *Nautical Almanac*—useful, but not entrancing, reading.

J. W. N. Sullivan

IN EARLY 1857 WILLIAM WALKER, a Tennessee native known to his admirers as "the grey-eyed man of destiny," found himself besieged with some 370 fellow countrymen in Rivas, a town on the Pacific coast of Nicaragua. He had boldly taken control of the country with his band of adventurers two years before, and his regime had been quickly recognized by the administration of US President Franklin Pierce. But Walker's decision to repeal Nicaragua's emancipation legislation of 1824, although exciting white supremacists with a vision of a great slave empire encompassing the entire Caribbean basin, roused opposition from neighboring Central American countries. These countries, aided by agents of Cornelius Vanderbilt, whose lucrative shipping business in the region Walker had imprudently interfered with, had sent the forces that had now placed Walker on the verge of annihilation.

In this moment of crisis, there providentially appeared a US naval vessel, the *St. Mary's*. Although this ship was primarily engaged in surveying for guano deposits, and its commander, Charles Henry Davis, had spent the previous seven years sailing a desk as head of the *Nautical Almanac* Office in Cambridge, Massachusetts, the *St. Mary's* presented a sufficiently martial aspect as to encourage the besieging force to enter negotiations. Davis was unsympathetic with Walker's goals, but he felt obliged to protect the lives of US citizens and therefore effected an armistice whereby Walker would give up his claim to Nicaragua, while he and the core members of his mercenary army would be allowed safe passage out of the country. The *St. Mary's* ferried the evacuees to Panama, for return to the United States.

When commander Davis at last returned from his sea duty in 1859, he found himself vilified by Walker's supporters in Congress and the press, who claimed that the man of destiny would have triumphed in Nicaragua had Davis not plotted behind his back to coerce his surrender. Davis shrugged off these criticisms and returned to his work as a scientific bureaucrat; moreover, he was gratified to learn that while he had been away his translation of a Latin text on celestial mechanics, by the great German mathematician and astronomer Carl Friedrich Gauss, had been published to some acclaim. This book marked the first rendering into English of the important data analysis technique known as the method of least squares.

Walker meanwhile soon broke his promise to cease meddling in Central American politics, but the US Navy was not present to save him when he next got in trouble. He was executed by firing squad in Honduras in September 1860.

══════

The William Walker adventure was only one of several dramatic, politically charged incidents that punctuated the long naval career of Charles Henry Davis. His life when stationed ashore, in contrast, was one of quiet administrative competence. He was a skilled mathematician, but he made his most significant mark by organizing the mathematical talents of others. In so doing he helped nurture a whole generation of mathematicians and mathematical astronomers who would influence higher mathematics in the United States for many years. He did not himself perform the calculations, nor did he provide the theory behind the calculations. But he played the indispensable role of bringing together the theorists and the calculators and making the whole operation run smoothly.

Davis was born into the social class that Nathaniel Bowditch success-fully attained only with great effort: the Boston elite. He attended the Boston Latin School, which had been founded in 1635 and survives to the present day, giving it claim to be the oldest school in the United States. In 1821, at age fourteen, Davis entered Harvard, founded in 1636. He did not graduate with his class in 1825, however. Instead, in 1823, he joined the US Navy as a midshipman, taking advantage of the influence of an uncle with connections in that service. At that time the navy, unlike the army, had not established a school to train its young officers. Instead, prospective offi-cers were mostly selected from a small collection of eminent families, among them the Davis clan, and then immediately subjected to a rigorous probationary period at sea.

By the summer of 1825 Davis found himself in his first great naval adventure. He was a junior officer aboard the schooner *Dolphin*, as that ship was dispatched to the western Pacific to bring to justice the muti-nous crew of the whale ship *Globe*. Whaling was a major economic activ-ity at that time, before the discovery of underground oil deposits, and disruptions in the industry could not be tolerated. The *Dolphin* ventured into areas previously little or never explored by men of European heritage, presenting the crew with extreme navigational and meteorological chal-lenges. They were thrust into encounters with native peoples whose cus-toms and languages they struggled to understand, sometimes resulting in violent conflict. Ultimately only two surviving mutineers were found, the others having perished at the hands of the natives or in clashes with each other. Davis's son, in his memoir of his father, held that this "early experi-ence in fighting and danger and adventure, and, above all, in responsibil-ity," had been character forming. It set him on the course to becoming "not only an accomplished officer, but a scholar and a student of science."

Davis returned to the United States in 1827, served aboard naval ves-sels in the eastern Atlantic and the Caribbean, and then in 1829 success-fully passed the examination that made him a permanent officer in the US Navy. For the next 11 years he participated in a variety of extensive cruises: he saw northern Europe from England to Russia; visited the Mediterranean from Gibraltar to Turkey; and made two more traverses of Cape Horn to and from Peru. During leisure time he became expert on the theory and practice of navigation, and he learned French, Spanish, and some Italian. In 1835 his father and sisters had taken up residence in Cambridge, Massachusetts, and in consequence this would become Davis's shore home for over three decades. Here he established a life-long friendship with Nathaniel Bowditch's protégé Benjamin Peirce, two years younger than Davis, who had become professor of mathematics and natural philosophy at Harvard at the age of twenty-four. Peirce would eventually guide Davis deeply into the study of mathematics.

In 1840 Davis found his naval career stifled. Opportunities for promotion were extremely limited, but at the same time younger men were now filling the shipboard duty stations. Living in Cambridge, waiting for something to turn up, Davis intensified his mathematical studies with Peirce and resumed work on the Harvard degree from which he had stepped away in 1823. He graduated in 1842. In that same year he married the sister of Benjamin Peirce's wife. The Peirce and Davis households were across the street from each other. Zoologist Louis Agassiz lived next door. Nearby were poets Henry Wadsworth Longfellow and James Russell Lowell, professor of Greek literature Cornelius Felton, and botanist Asa Gray. The Cambridge of this era, as Davis's son would later describe it, was populated by "names which belong to the front rank in the intellectual life of the country."

Also in 1842 Davis joined the Coast Survey, for which Peirce was already acting as a consultant. Here Davis remained until 1849, and during this period he became a close friend of the director, Alexander Dallas Bache, who had assumed this position in 1843. This friendship would prove to be of immense value to Davis, as Bache was one of the most politically astute and well-connected American scientists of that or any era: a West Point graduate, a great-grandson of Benjamin Franklin, and a nephew of George Mifflin Dallas, vice president of the United States from 1845 to 1849. Bache used these advantages to revitalize the Coast Survey, and more broadly to create a small but influential place for science in the federal government.

Thus by the end of the 1840s, through his friendships with Peirce and Bache, Davis had become intimately connected with the two most substantial centers of scientific activity in the United States: Cambridge, Massachusetts, and Washington, DC.

At the Coast Survey, Davis became expert on the action of the tides and ocean currents on the surface of the seafloor. He published two learned disquisitions on this subject, based on his own observations and his reading in other authorities. The practical import of his work was felt especially with regard to the Nantucket Shoals, where production of accurate charts and the building of a new lighthouse greatly decreased the dangers for vessels of all sorts.

In 1849 Davis's demonstrated skills and his political connections gained him appointment as superintendent of a project to publish an *American Ephemeris and Nautical Almanac*, under the jurisdiction of the secretary of the navy. The aim was to produce, on a regular basis, accurate tables of the positions of the sun, moon, and planets that could be used by sailors both naval and commercial. Such a work was a necessary companion to Bowditch's *Practical Navigator*, in which the reader is frequently advised to consult the "Nautical Almanac" for the data needed to carry out various shipboard calculations. The British had been producing such a publication for many years, but the United States sought to supersede that work.

The basic problem was to describe the position of a heavenly body—say, the planet Jupiter—as it traverses the sky. The stars, as observers had noted many thousands of years ago, can be usefully thought of as affixed to a gigantic sphere, rotating around the earth. In the Northern Hemisphere, it was further noted that there was one star that never moved. This North Star could be taken as one pole of the axis of rotation of the celestial sphere, with the celestial equator conceived of as the circumference of a disk perpendicular to that axis, and at 90° from the North Pole. In other words, the celestial equator is a huge circle exactly parallel to the earth's equator. While the stars appeared to move in precise circles around the axis, never changing their positions relative to each other, Jupiter was among a small group of superficially starlike objects called planets (five for the ancients, seven for mid-nineteenth-century astronomers, aided by telescopes) that moved in more complicated fashion.

The position of Jupiter with respect to the stars of the celestial sphere, at any given time, can be given by two numbers, in analogy with latitude and longitude on the surface of the earth. In the *Nautical Almanac* these two numbers are the declination (corresponding to latitude), measured in degrees, from 0° at the equator to 90° at the North Pole of the celestial sphere; and the right ascension, measured in hours around the celestial equator, from 0 to 24. As with longitude, British naval supremacy had established a tradition of setting the zero of right ascension in accordance with the location of the Greenwich Observatory, outside London. In the United States there was nationalistic feeling that deference should be given to the Naval Observatory in Washington. The American *Nautical Almanac* of the Davis era resolved the issue by providing two sets of tables.

Ancient observers had compiled considerable data, essentially equivalent to declination and right ascension, describing the paths of Jupiter and the other planets over time, and had proposed a variety of cumbersome explanatory theories. Nineteenth-century scientists had greatly improved the data, thanks to the telescope, and moreover had developed a much more coherent theory. As noted in chapter 1, a small set of basic laws of motion, and the principle of universal gravitation, together with the resources of the calculus, were capable of yielding differential equations to represent the motions, not only of Jupiter but of all the other planets, and of the sun and moon as well.

Looked at in broad outline this solution was satisfyingly elegant. But if one wished to produce precise numbers for the position of Jupiter for every hour of every day of some future year, as was needed for the *Nautical Almanac*, the details were not so aesthetically pleasing. The solutions to the differential equations of celestial mechanics were in general not simple formulas, because each planet is not traveling around the sun in isolation but is interacting with the other planets in a complicated dance of gravitational forces. Faced with this situation, astronomers and mathematicians developed schemes for approximating the solutions of the differential equations, schemes they were continually revising. All required, to produce the actual coordinates of a planet, a substantial series of numerical computations.

Davis's task, as superintendent, was to organize a team to accomplish these computations. He was allowed to set up the *Nautical Almanac* Office to suit his own convenience, in Cambridge. Relying on observational data from around the world, with special advice from the astronomers at the Harvard Observatory, he hired Benjamin Peirce as a consultant to oversee the theoretical foundations, and he assembled a collection of "computers" to perform the arithmetic, all by hand. As noted, the computations needed to be redone, year after year. The first edition of the *American Ephemeris and Nautical Almanac* was published in 1852, providing numbers for 1855; in 1853 the 1856 numbers were published, and so on.

Working as a computer was not in itself intellectually challenging. Simon Newcomb, who began work at the *Nautical Almanac* Office in 1857 at age twenty-two, while Davis was away commanding the *St. Mary's*, was surprised and disappointed to find that his duties required no real knowledge of the mathematical intricacies of celestial mechanics. Davis and Peirce had carefully designed the system so that all Newcomb needed to know was arithmetic and the ability to follow directions. Nevertheless, the opportunity to serve as a *Nautical Almanac* computer attracted individuals of high mathematical competence and produced a stimulating intellectual environment. The *Nautical Almanac* helped launch several notable scientific careers, including Newcomb's. He would become a prominent mathematical astronomer and would himself be appointed superintendent of the *Nautical Almanac*. Other computers in the early years were John Runkle, who became president of the Massachusetts Institute of Technology; John Monroe Van Vleck, who became a professor of mathematics and astronomy at Wesleyan University; Maria Mitchell, who became a professor of astronomy at Vassar College and director of its observatory; and T. H. Safford, who became a professor of mathematics and astronomy at Williams College.

Once his computational system was running smoothly, Davis embarked on the translation project referenced at the beginning of the chapter. He was combining his classical education and his knowledge of advanced mathematics to translate from Latin to English a work of eminent usefulness to the *Nautical Almanac* mission: *Theoria Motus Corporum Coelestium in Sectionibus Conicis Solem Ambientium.* Carl Friedrich Gauss, author of the monograph, had burst upon the field of mathematical astronomy in 1801, at age twenty-four, by correctly predicting the position of the tiny asteroid Ceres, after it had been briefly observed by an Italian astronomer earlier in the year, and then lost to telescopic sight. In *Theoria Motus*, originally published in 1809, Gauss described the secrets of his success, demonstrating rigorous methods for approximating the orbit of a celestial body on the basis of a few observations. In particular, Gauss showed the utility of the approach that became known as the method of least squares. Suppose one wishes to choose the best mathematical model fitting a set of observed data from among a family of potential mathematical models, say, elliptical orbits of a planet. The idea is to compare the deviations between the observed data and each of the hypothesized mathematical models, and then to choose the particular model that minimizes the sum of the squares of the deviations. The squaring operation has the effect of giving equal weight to both positive and negative deviations. The French mathematician Adrien-Marie Legendre had published the basics of the method in 1805, but the famously precocious Gauss had claimed, plausibly, that he had discovered it as a teenager in the 1790s.

Latin for many years had been the favored language of European scholarship, but the German-speaking Gauss was one of the last major mathematicians to persist in this tradition. French and German would dominate such writing in the nineteenth century, to be supplanted by English in the twentieth.

As noted previously, Davis's translation, *Theory of the Motion of the Heavenly Bodies Moving about the Sun in Conic Sections,* was published in 1857, while he was out of the country on the *St. Mary's.* He had taken command of that ship because he saw it as his last chance for advancement in the navy. The alternative of remaining in charge of the *Nautical Almanac* indefinitely, but with no chance for promotion, did not appeal. Despite his scientific success onshore, he always chose to serve afloat when given an opportunity, frequently entangling himself in political controversy thereby.

Davis returned to Cambridge from his cruise in 1859 and immediately resumed superintendence of the *Nautical Almanac,* but in the crisis atmosphere of 1861 he was called to Washington. The navy was under extreme organizational stress, with large numbers of officers leaving the service in consequence of the secession of the Southern states. Davis was assigned to help manage the assignment of officers and the purchase of ships. He was also placed on a commission given the task of planning a blockade and other naval operations against the South. When hostilities broke out, he was thus well placed to witness, and to help implement, the brilliant bureaucratic maneuvering of his old friend Bache with regard to the Coast Survey. While others in government were predicting, even relishing, the demise of the survey as a luxury unaffordable in time of war, Bache quietly succeeded in making it indispensable to the war effort. He made sure that the War Department understood the value of the maps already produced by the survey, and he attached his assistants to the forces in the field as topographic experts, both for the coast and the inland waterways.

At first Davis combined his Washington duties with continued supervision of the *Nautical Almanac*, by correspondence, but later in 1861 he gave up both to take a more direct role in the war effort. He was one of the principal planners and participants in the successful capture of Port Royal, on the coast of South Carolina, after which he was promoted to captain. Then in 1862 he had his first experience of freshwater naval operations, where he was instrumental in the capture of Memphis and in other efforts of the Union forces to secure the Mississippi River. This campaign would have been unimaginable without steam-powered vessels.

Davis's son later wrote glowingly of his father's performance on the Mississippi, but Civil War historian Shelby Foote, writing in the 1950s, dismissed Davis as an aristocratic intellectual who lacked the aggressiveness needed for command in a combat zone. Davis's subsequent recall to Washington, and his promotion to rear admiral as head of the newly founded Bureau of Navigation, overseeing all the scientific activities of the navy, Foote saw as a classic bureaucratic ploy to kick Davis upstairs.

By this time Davis had become, along with his old friends Agassiz, Bache, and Peirce, a core member of a group calling themselves the Scientific Lazzaroni, in ironic reference to the homeless beggars of Naples. At first this American clique was little more than an informal social club of

Charles Henry Davis, 1861–70. Courtesy of Special Collections, Fine Arts Library, Harvard University

convivial like-minded intellectuals. They were in no sense homeless, but they did sometimes feel underappreciated, forced to solicit funds and recognition commensurate with their lofty view of their talents. In 1863 they persuaded Senator Henry Wilson of Massachusetts to push a bill through Congress creating the National Academy of Sciences. Soon signed into law by President Lincoln, this was to be a counterpart to the Royal Society in England, and moreover was to advise the government on scientific matters. An initial membership of 50 was created, including Davis and his three friends. Jealousies regarding who was in and who was out placed great strain upon the academy in its first years of existence, but it has survived to the present day as an honorary society of scientists, although its governmental advisory role did not come to fruition until the twentieth century.

After the war Davis became superintendent of the Naval Observatory. He served there until 1867, when he was named commander of the navy's

Brazil squadron. Here he would have his last shipboard adventure. Upon reporting to the squadron, Davis found himself immediately embroiled in an ongoing conflict, known as the War of the Triple Alliance, pitting Paraguay, remotely situated in the interior of the South American continent, against the three coastal nations of Argentina, Brazil, and Uruguay. Once again Davis was called upon to rescue Americans from perilous circumstances that, Davis strongly suspected, were the consequence of their own stupidity or malfeasance. In this case the individuals in dire straits were the US minister to Paraguay, Charles Washburn, and two associates who had been under the protection of the US legation, though not officially employed by the government. Because communication with Washington was erratic and subject to long delays, Davis was forced to make decisions on his own amid extremely confusing conditions. The United States at the time was attempting to maintain neutrality among the several nations concerned, while accusations were volleyed back and forth about torture tactics applied by Paraguay's leaders and double dealing by American diplomats. Davis managed to bring all the threatened individuals to safety, only to have them, as in the case of Walker in the 1850s, complain afterward about discourteous treatment and usurped authority. A congressional inquiry was initiated in 1869, during which Davis's conduct was sharply questioned. His naval colleagues came stoutly to his defense, and no formal action was taken against him. But Davis's son, himself a naval officer, long nurtured a grievance, feeling that his father had in essence been persecuted for his high birth and cultivated background. As if to confirm his patrician status, in the same year that Davis was being grilled by Congress, Harvard awarded him an honorary doctor of laws degree.

After the Paraguay hearings, Davis commanded the naval station at Norfolk, Virginia, for three years and then served another tour as superintendent of the Naval Observatory in Washington. He chaired the Transit of Venus Commission of 1874, helped prepare the observatory exhibit at the 1876 Centennial Exposition in Philadelphia, and edited the journals and memoranda of the Arctic voyage of the *Polaris*. Admiral Davis died early in 1877.

7

General Principles

Daniel Harvey Hill

The operations of the mind, such as hope, fear, joy, grief, &c., are not
quantities. For, although we speak of a great hope and a small hope, there
is no definite unit of comparison by which to measure its magnitude.

D. H. Hill

EARLY ON THE MORNING of September 14, 1862, on the slope of South
Mountain, in central Maryland, two army officers on horseback were
making a cautious reconnaissance along a wooded track when they came
upon a man and his young children, standing apprehensively in front of a
cabin. The father, glimpsing the blue cloak of the younger officer in the
gloom of the overhanging trees, volunteered the location of the "rebel"
forces. "The road on which *your* battery is," he explained, "comes into the
valley road near the church." Pleased to receive this important piece of intelligence, the officers were preparing to depart, when an artillery shell
crashed through the woods and one of the children began to cry. The senior officer, thinking fondly of his own little daughter back in North Carolina, spoke comforting words to the frightened child. Then the two men resumed their way, with the junior officer grateful, on this cold morning, for
the cloak he had picked up from a fallen enemy soldier on a Virginia battlefield earlier that year.

The older officer was Daniel Harvey Hill, less than two years removed
from being a college professor of mathematics but now a major general in
the army of the Confederate States of America. The other officer had once
been Hill's student. The two men were part of the invasion of Maryland by

Confederate forces, which would culminate three days later in one of the most violent confrontations of the Civil War, the Battle of Antietam.

═══

Harvey Hill's professional rise in both soldiering and mathematics began in the same place: the US Military Academy at West Point, New York, from which he graduated in 1842. Born on a plantation in South Carolina, with a family tradition of military service, including two grandfathers who were veterans of the Revolutionary War, the teenage Hill had had no notion when he traveled north that he was entering the country's leading institution for advanced mathematics instruction.

As an entering student, Hill felt himself at a marked disadvantage with the Yankee boys, many of whom had already had a stiff dose of arithmetic, and even some algebra and geometry, while Hill in the South received a literary education that emphasized Latin and Greek. But he persevered with his mathematical studies at the Point, so that despite not achieving the highest grades in the subject, he felt confident in his abilities by the time he graduated.

Hill stayed in the army through the war with Mexico of 1846–48, establishing himself as brave under fire and hot tempered in the face of the incompetence he so often detected. Because he published the harshest criticisms of his superiors anonymously, he avoided lasting damage to his reputation. This was well, since the officer corps in those days was a compact fraternity; everyone knew everyone else. Thus Hill in the 1840s encountered a large proportion of the men who would figure prominently in the 1860s, on both sides of the War of the Rebellion. He formed decided judgments of many of them but was not always prescient. The emergence of Ulysses S. Grant (West Point, 1843) was a great surprise. "Grant was not once thought of," Hill later admitted, when he reflected on his 1861 assessment of who the South had most to fear among the North's military leaders.

The Mexican War had greatly intensified the sectional tensions that had been simmering for decades. Hill questioned the conduct of the war, but not the wisdom of going to war in the first place. Grant, like many others

in the North, saw the run-up to the war as "a conspiracy to acquire territory out of which slave states might be formed for the American Union,"
and the war itself as "one of the most unjust ever waged by a stronger
against a weaker nation." Still, like Hill, he went to the war when ordered
and served loyally.

The war derailed Grant's plan to be a college mathematics professor,
an aspiration to which he never returned. Not so with Hill, who in 1849 resigned from the service to take a mathematics professorship at Washington College in Lexington, Virginia, followed by a similar position at Davidson College, outside Charlotte, North Carolina. In those years Hill and his
wife produced nine children, burying four before adulthood. Three of his
own siblings had died prematurely, so Hill was well acquainted with the fra-
gility of a nineteenth-century child's life, but he never got used to it. He
was always sensitive to the vulnerability of children, as he displayed that
morning on South Mountain.

Harvey Hill was no great mathematician. In the Military Academy's
well-stocked library he had had access to works by most of the major figures of European science and mathematics: from the ancient Greeks such
as Euclid and Apollonius; through Galileo, Huygens, Newton, and other
giants of the seventeenth-century scientific revolution; and on to research
monographs by the titans of the then recent past, including Euler, Lagrange, Laplace, and Legendre. But Hill showed no inclination or talent to
contribute to the leading edge of mathematical knowledge, pure or applied.
Vanishingly few in the United States did in that era, and this would not
change until much later in the nineteenth century. Nevertheless, Hill did
take a stab at the growing field of textbook writing.

Hill's *Elements of Algebra* of 1857 began with a set of testimonials, led
by one from his brother-in-law Thomas J. Jackson, another mathematics
professor trained at West Point (class of 1846). It was a generally competent book, although he did mangle some key subtleties. For instance, Hill
did not grasp the importance of the fact that all whole numbers can be factored into primes in only one way, aside from the order of the factors. And
Hill, like many textbook writers of the time, was leery of square roots of
negative numbers, so-called imaginary numbers. He did not accept the

growing consensus of leading mathematicians that these numbers were an unremarkable convenience.

There were books similar to Hill's on the market, some with exactly the same title, and the West Point library contained several of them. Did Hill borrow from these books in writing his own? Few would have known or cared. Observance of copyright in the United States was casual at that time, especially for foreign books. But no, although naturally sharing many features with other introductory algebra books, Hill's book shows no evidence of plagiarism. In some respects it was strikingly original, an originality at the heart of how D. H. Hill came to lead a contingent of soldiers in Maryland in 1862, intent on killing and maiming their fellow Americans.

Hill's algebra book reveals a man who had been building a list of grievances through the 1850s. He had abiding friendships with some individuals from the North, notably some of his West Point classmates, but overall he despised Yankees, especially those self-righteous New Englanders, looking down their noses at the immorality of slaveholding while being up to their necks in the triangle trade, and even now happy to grab profits deriving from the slave economy while avoiding the day-to-day reality of holding people in bondage. Hill used the exercises in his book, what would now be called story problems or word problems, to depict Northerners as cheaters in commercial transactions, cowardly in the face of danger, tolerant of the absurd notion of women's rights, and hypocritically miserly when given the opportunity to buy the freedom of a slave. Other problems from the text highlighted the Northern slaughter and enslavement of the Indians, and the despicable Salem witch trials. Hill also pointed up the disloyalty of the New England states during the War of 1812, hoping that readers would note the similarity to the more recent defiance of the federal government regarding the Fugitive Slave Law. All this in support of teaching geometric proportion, simultaneous linear equations, and quadratic equations.

Hill in his algebra book flaunted slavery, without embarrassment, as integral to the Southern way of life. A farmer buys land and slaves together, and the relative worth of each must be ascertained. A planter hires a slave and the slave's clothing at a certain annual rate and then returns the slave to his master after only eight months, with a cash payment but minus the clothes. What was the value of the clothes? If a man can do a piece of work in five days when the days are 12 hours long, how long will it take him when the days are 15 hours long? A bit of proportional reasoning gives the answer of four days. But can a man be induced to work outside in the hot sun for 12 consecutive hours, let alone 15? Certainly, if the worker is an enslaved "negro-man," as is assumed in Hill's book.

Despite his animus against the North, Hill had not sought war, he had not recommended it as a wise policy for the South, but he had feared that war was coming. He knew the North's capability and had no rosy view of Southern prowess at arms. Thus in 1858, hoping to increase Southern preparedness, he resigned from Davidson to become superintendent and professor of mathematics and artillery at the newly founded North Carolina Military Institute in Charlotte. When hostilities began, Hill led the entire body of the college, students and faculty, into the Confederate service. He remained loyal to the cause through four long years of war, despite witnessing some of the worst of the slaughter and often clashing with the political and military leaders under whom he served. Even Robert E. Lee was no hero for Hill. Lee's Pennsylvania invasion of 1863 seemed to Hill a grave mistake. Hill retained to the end of his life the belief that better leadership could have saved the South as an independent nation.

In the war Professor Thomas J. Jackson was transformed into General Stonewall Jackson and became Lee's most trusted subordinate, before being accidentally killed by his own men in 1863. Ulysses S. Grant, mediocre farmer and businessman, emerged as the overall commander of the Union armies and led them in crushing the rebels. Harvey Hill survived the war, returned to his family in North Carolina, and became a journalist. From this position he was able to continue attacking the hypocrisy of the North and to express his outrage at the treatment of the South, especially the attempt to give civil rights to the newly freed population. He scoffed at the attempts of the freedmen to contribute to government. Yet he considered himself a deeply devout Christian and a custodian of *The Land We Love*, as he titled the periodical he edited in Charlotte from 1866 to 1869. Not that he thought the South was, or ever had been, perfect. He came to the realization that slavery had been an evil—for white Southerners. It had been

Major General Daniel Harvey Hill. Cook Collection, The Valentine, Richmond, Virginia

responsible for the South's backwardness in science and technology. The North, Hill observed, had been motivated to invent and promote mechanical devices, and to invest in scientific knowledge, to make up for the lack of an enslaved population. The Southern ruling class, in contrast, with large quantities of workers always at command, ignored technological advancement while developing refined manners and great political skills, with which they long dominated the national government. But those skillful Southern politicians had been of no help when it came to the brutal reality of the war. Hill's sociological analysis incidentally explained his poor preparation in mathematics compared to his Yankee classmates. According to Hill, the North, unlike the South, had understood that mathematics was "the essential pre-requisite" to the mastery of science and technology. Now that slavery was gone, it was time for the South to catch up educationally. He recognized that this would be difficult, given the attitudes of Southern educational leaders:

About the same time the eminent President of a Southern college delivered and published an address to prove that the standard of mathematical science in our institutions of learning ought to be lowered. (Until then we had supposed that zero was the lowest figure in the table of numbers.) The system of instruction proposed by this great, good, and wise man was no doubt adapted to make profound thinkers on abstruse and metaphysical points; but it could never have made one single practical and useful man.

As Reconstruction wound down following the presidential election of 1876, Hill's blistering anti-Northern rhetoric cooled. With white Southerners increasingly able to dominate their black neighbors through legal and extralegal means, without federal interference, he began to talk of reconciliation. He continued, in magazine articles, to refight some of the controversial moments of the war, but he now had kind words for commanders on both sides. He predicted that "waves of oblivion will roll over the bitter recollection of the strife."

Seeking to realize the educational improvements he had advocated as a journalist, Hill returned to academe, first as president of the Arkansas Industrial University (which became the University of Arkansas) and then as president of the Middle Georgia Military and Agricultural College (later Georgia Military College). All this pedagogical effort was on behalf of white Southerners only. Hill was silent on educating the formerly enslaved and their descendants.

Hill died in 1889. Great changes had indeed occurred since the national trauma of the 1860s. Antietam had become the site of a national cemetery, containing more than 4,700 war dead. The Military Academy had continued to emphasize mathematics, but its role as a major national incubator of mathematics teachers had faded. Algebra textbooks had continued to be written with story problems, but they were comfortably bland. Stonewall Jackson had been deified by supporters of the lost cause of Southern independence. Ulysses S. Grant had served two terms as president and died in 1885. But the waves foreseen by Harvey Hill regarding the Civil War have been not so much waves of oblivion as waves of ignorance and obfuscation.

8

Fellow Worker

Christine Ladd-Franklin

1878

I have written to Miss Ladd saying that I did not personally anticipate that
her sex would be an objection when attending lectures at our University
and I should rejoice to have her as a fellow worker among us.

J. J. Sylvester

I N THE FALL OF 1878, Christine Ladd, a thirty-one-year-old country
schoolteacher, began attending graduate mathematics courses at Johns
Hopkins University in Baltimore. Although paying no tuition, she participated fully in the classes and so impressed the professors as to be awarded
a fellowship for the next academic year. By 1882 Ladd had satisfied all the
requirements for a PhD in mathematics, culminating in a dissertation titled "On the Algebra of Logic," written under the supervision of one of the
greatest minds of the nineteenth century. The university refused to award
this degree until 1926, when Ladd was seventy-eight years old.

Hopkins had opened its doors only two years before Ladd's entry but
had already established itself as a preeminent center of advanced education in the entire country. Indeed, the excitement about Hopkins in scholarly circles had begun even before the students arrived, as word spread
about the generous salaries being offered to the faculty, and as it became
clear that the new institution intended to emulate the German universities
by emphasizing postgraduate study and original research. The new university's first president, Daniel Coit Gilman, had no special affinity nor deep
acquaintance with mathematics, but by choosing the extroverted and inspiring James Joseph Sylvester as the first chairman of the mathematics department, Gilman ensured that the subject would not go unnoticed at

Hopkins. The selection of Sylvester also incidentally made it possible for an unconventional student such as Ladd to be given an opportunity at the university.

≡≡≡≡

Christine Ladd had been born in 1847, and with her parents' encouragement she developed an early thirst for learning while attending a variety of schools in New England. Her mother had been an activist for women's rights but died when her daughter was twelve. Christine's father continued to support his daughter's aspirations, although his health problems and peripatetic merchant career occasionally interfered with Christine's educational advancement. She was excited to read about the founding of Vassar, a college for women in Poughkeepsie, New York, and persuaded her father and grandmother to allow her to enroll there in 1866, the year after its founding.

Vassar was part of a wave of institutional innovation after the Civil War that expanded opportunities for women to pursue collegiate education. This era saw the founding of Smith (1871), Wellesley (1875), and Bryn Mawr (1885) as women's colleges, along with Cornell (coeducational from its founding in 1865). At Vassar, Ladd displayed wide-ranging scientific interests in physics, chemistry, astronomy, and mathematics, and was especially energized by the astronomer Maria Mitchell, by far the most prominent woman scientist in the nineteenth-century United States. Mitchell had achieved international recognition by discovering a new comet in 1847. As noted in chapter 6, Mitchell had been a computer for the *Nautical Almanac*, before joining the faculty at Vassar.

After graduating from Vassar in 1869, Ladd spent the next few years teaching in a variety of secondary schools, never becoming rooted in any one place. Whenever the opportunity presented itself, she furthered her own education, attending mathematics lectures at Washington and Jefferson College in Pennsylvania, Cornell in New York, and Harvard in Massachusetts. Only at Cornell would she have been able to enroll officially for a degree, but it did not yet offer the advanced studies for which she had begun to hunger. Johns Hopkins University emerged just as she was pondering her next move, attracting her attention by offering the rigorous research emphasis she desired.

The question of enrolling women bedeviled the founding administrators of Johns Hopkins University and continued to perplex their successors for nearly a century. The era of the founding, as noted, was one in which women were increasing their visibility in higher education, and so the Hopkins trustees were exposed to emphatic arguments for making the institution coeducational. Some of these arguments came from their own daughters. But full adherence to the cherished German model, as it existed at the time, suggested a complete prohibition of women students. There were also thoughts of a middle way, a separate coordinate college for women. This was to be the solution chosen by Harvard (with Radcliffe) and Columbia (with Barnard). The Women's College of Baltimore (later renamed Goucher College), founded in 1888, was not officially connected with Johns Hopkins, but was considered by many as de facto coordination and thus served to assist the conservatives in keeping the undergraduate population of Hopkins all male until 1970.

But at the graduate level there proved to be a small opening for women, as Christine Ladd demonstrated. Her approach to the university was to appeal directly to mathematics department chairman Sylvester. He was no stranger to institutional prejudices; as a Jew in his native England he had long been denied a position consonant with his accomplishments. Ladd's own accomplishments were modest, but she was able to claim credit for not only her Vassar degree but also a series of short articles she had published in mathematical journals.

These were not groundbreaking works. One of Ladd's first publications was a mere summary of the contents of a recent issue of a European mathematics journal, which, she noted with disapproval, "lies on the shelves of the Boston Public Library with uncut leaves." In another short piece, a "resumé" of "the quaternion equivalents for all the transformations of Trigonometry," Ladd was so profuse with symbolic notation that the printer apologetically announced in a footnote that he had run out of capital U and been forced to substitute the Greek letter upsilon. But these efforts of Ladd demonstrated a keen interest in mathematics and, most important for Sylvester, an awareness that it was not static but live and growing. Sylvester became convinced that she had the aptitude and background needed for graduate study and argued her case with the university administration. He was asserting his right of *Lehrfreiheit*, which in Germany expressed the freedom enjoyed by professors at the forefront of knowledge to teach whatever and however they wanted. When Ladd proved worthy of Sylvester's regard, other mathematics professors at Hopkins followed his lead and let Ladd into their classes as well.

After a year, the trustees approved her continued attendance and voted a stipend. This decision was announced in the *Johns Hopkins University Circulars*, published in December 1879, with the supporting comment that "Miss Christine Ladd" was "a graduate of Vassar College, whose mathematical attainments had been commended by the Faculty as worthy of the holder of a Fellowship in Mathematics." She was the only fellowship holder for whom such an explanation of worthiness was offered. The announcement was repeated in January 1881.

Ladd's activities at Hopkins were reported in the *Circulars* just like any other student, except that her place of residence was not listed with those of other graduate students, and Vassar College was not included in the "List of Institutions at which Graduate Students received their Bachelor's Degree." Her publications were touted; her enrollment in mathematics and logic courses was noted; the papers she presented at meetings of the Metaphysical Club were recorded.

The Metaphysical Club was one of several university societies or associations encouraged by President Gilman as venues for presentation of research reports, by both faculty and students, across disciplinary boundaries. The meetings of these groups were designed to be less formal and less specialized than the departmental "seminaries" that Gilman also encouraged, one of his favorite imports from Germany. In addition to the Metaphysical Club, the early Hopkins era also featured the Philological Association, the Scientific Association, the History and Political Science Association, the Biological Association, and the Naturalists Field Club. These groups usually met once a month. Ladd's attachment to the

Metaphysical Club shows the influence of its founder, Charles Sanders Peirce. Ladd owed her access to Hopkins to Sylvester and fondly recalled his "boyish enthusiasm" (he was sixty-four when they first met), but it was the "brooding" Peirce who made the deepest imprint.

Peirce, the most brilliant of the sons of Harvard mathematician Benjamin Peirce, had been born in 1839 and raised up in the stimulating intellectual environment of Cambridge. Here there had been an earlier Metaphysical Club, where the younger Peirce, according to his friend William James, had played the central role in birthing the distinctively American philosophy known as pragmatism. Unlike James, Peirce never produced provocative but accessible books encapsulating his thoughts. Instead, Peirce scattered his often obscure cogitations in small chunks across a variety of journals: *Popular Science Monthly*, *Journal of Speculative Philosophy*, *North American Review*, *The Monist*, and *Educational Review*, among others. He left a large quantity of unpublished writings. Moreover, Peirce's interests were immensely various. In addition to subtle pondering on the nature of science and on human knowledge in general, he made careful analyses of the technicalities of measurement associated with the day-to-day operation of the Coast Survey, where he worked for many years. He delved into aspects of the most abstruse notions of the pure mathematics of his day while also contributing to linguistics, psychology, and economics. His genius has been regularly proclaimed since his death.

Peirce had only a modest scholarly reputation when he came to Hopkins in 1879. While retaining his position with the Coast Survey, he signed on as a part-time lecturer in logic, a field to which he had been giving considerable attention. He was among the few who had recognized how thoroughly logic had been revolutionized by the recent work of the English mathematicians George Boole, Augustus De Morgan, and their followers. They had broken the 2,000-year grip of Aristotle, moving decisively beyond the classic syllogisms (All men are mortal; Socrates is a man; therefore . . .) to a more comprehensive view of logical relationships. By designating propositions with letters and treating the logical operations of *and*, *or*, and *not* analogously with the operations of multiplication, addition, and negation in arithmetic, they turned logical deduction into an exercise in rule-based symbol manipulation, like algebra. The details and the implications took decades to work out, but logic had effectively become a part of mathematics and not an odd appendage of philosophy. Charles Peirce's insight into this field, rich with research opportunities, inspired several of his Hopkins students. Christine Ladd was one.

In 1882 Peirce collected the work of his four best students, together with a contribution of his own, and arranged for publication, under the title *Studies in Logic by Members of the Johns Hopkins University.* He lauded Ladd's piece "On the Algebra of Logic" as serving "for those who are unacquainted with Boole's *Laws of Thought,* as an introduction to the most wonderful and fecund discovery of modern logic." He also noted that "Miss Ladd" was "now Mrs. Fabian Franklin."

Franklin, six years younger than Ladd, had come to Johns Hopkins as a graduate student in mathematics in 1877, after a brief career as a civil engineer. He and Ladd attended some of the same classes; they studied the theory of numbers with Sylvester, and they both took advanced logic from Peirce. The university conferred a PhD on Franklin in 1880, and then appointed him to a faculty position. Ladd was one of his students in modern algebra in the spring of 1882, and that same term they were both enrolled in algebraic geometry and abelian and theta functions, taught by the celebrated Arthur Cayley, visiting from England. They were married in August of that year.

Christine Ladd-Franklin (this was the authorial identification she would ultimately settle on, although she also used the unhyphenated form for some years), had a son in 1883 and a daughter in 1884. Only the daughter, who would in 1913 publish a book supporting women's suffrage, survived infancy. Ladd-Franklin enrolled in no classes after 1882, but she did some private tutoring in Baltimore and remained visible in the Hopkins community. In the early 1890s she proposed a lecture series on logic at the university but was rebuffed. Since her husband was on the faculty, an unwritten anti-nepotism policy may have been invoked, a ploy that universities often wielded against scholarly married women even into the second half of the twentieth century. Hopkins did at last hire Ladd-Franklin, with minimal compensation, as a part-time lecturer in logic and psychology in

1904, the only woman on the faculty. By that time her husband had left Hopkins to become a journalist. When Fabian Franklin became an associate editor of the *New York Evening Post* in 1909, the couple moved to New York City, and Christine lectured at Columbia University, unpaid.

Ladd-Franklin did not allow the foot dragging of university administrators to prevent her from pursuing her research interests. She gradually eased out of research in logic, although she kept up with developments. She was unafraid to offer candid critiques of the younger luminaries in the field, such as Giuseppe Peano in Italy and Bertrand Russell in England. In 1887 she became intrigued with some optical illusions related to the geometric aspects of binocular vision, and this interest blossomed into a research program on other features of vision, especially color vision. She would publish widely on this subject over the next 30 years. Henry P. Bowditch (the navigator's grandson), in a physiology textbook published in 1896, made favorable mention of Ladd-Franklin's theories of color vision.

In the 1890s Ladd-Franklin spent time in Göttingen and Berlin working with leading German vision investigators, during which she learned firsthand about the struggles in Germany, parallel to those in the United States, to open the universities to women. With her experiences in both countries, she became a slashing polemicist on feminist issues. In 1896 she skewered George Romanes, a disciple of Charles Darwin, for suggesting that women with intellectual ambitions should be pitied, because their almost certain failure would bring misery on both themselves and their friends and family: "But this theory of Mr. Romanes is one which does not need confirmation by facts. It is one of those theories which the strong intuitive powers of his sex can perceive to be true at a glance, and to which the dicta of experience are absolutely immaterial." Ladd-Franklin offered in response the accomplishments, and the sociable nature, of notable women intellectuals of the nineteenth century: Sonya Kovalevsky, Maria Mitchell, and Mary Somerville. Christine Ladd-Franklin saw herself in a great tradition, which she was attempting to expand.

Despite the obstacles in her path, the career of Christine Ladd-Franklin was in some ways more fortunate than that of her mentor, C. S. Peirce. He left Hopkins in 1884, and it would prove to be the only academic position

Christine Ladd-Franklin, ca. 1909. Johns Hopkins University Sheridan Libraries

he would ever have, aside from giving some guest lectures. Historians would later discover that he was dismissed from Hopkins largely because of charges of moral impropriety brought by his colleague, astronomer Simon Newcomb. Specifically, Peirce was shunned for having lived with his future wife before their marriage. His later years would be shadowed by poverty and vague suspicions of drug use and other irregular behavior. Daniel Coit Gilman came to consider Peirce such a reprobate that he once refused to stay under the same roof with him. Christine Ladd-Franklin, whose roof it was, remained supportive to the end, and she wrote a moving tribute after Peirce's death in 1916. She regretted the impenetrability of his later work, blaming his dismissal from Hopkins for cutting him off from the healthy give-and-take with bright and curious students that had invigorated his thought when she had first known him. In contrast to those who cast aspersions upon him, she lauded the "ingrained sweetness" of his character.

Ladd-Franklin should have received a PhD from Johns Hopkins University in 1882. But although the Board of Trustees had tolerated her participation in classes, conferring a degree was a step too far for them at that time. It was not until 1893 that the university conferred a doctorate on a woman, Florence Bascom in geology. Like Ladd, she had been admitted as a special student to avoid setting a precedent. In 1907 the board at last allowed women to enter as full-fledged graduate students. Even then, individual professors retained the right to bar women from their classes. The first mathematics PhD for a woman at Johns Hopkins was awarded to Clara Latimer Bacon in 1911.

In 1926 the Hopkins administration conceived a plan to award Ladd-Franklin an honorary degree, in connection with the ceremonies marking the fiftieth anniversary of the university. Ladd-Franklin, however, successfully insisted that rather than an honorary degree she be awarded the PhD rightfully owed her from 1882.

Although Ladd-Franklin would not forget the administrative intransigence at Johns Hopkins, she was deeply appreciative of the intellectual bounty that the university had presented her. She would forever remember the "ardor of research" and the "joy of the intellectual life" that she had shared at the early Hopkins.

9

Straddler

Kelly Miller

1887

The negro should plant one foot on the Ten Commandments and the other on the Binomial Theorem: he can then stand steadfast and immovable, however the rain of racial wrath may fall or the angry winds of prejudice may blow and beat upon him.

Kelly Miller

I N 1887 KELLY MILLER APPLIED to take graduate-level classes in mathematics, physics, and astronomy at Johns Hopkins University in Baltimore. He had been engaged in private study of mathematics at the US Naval Observatory in Washington, after graduating from college in 1886 with an excellent academic record. In addition, Miller was the son of a man who had served in the Confederate Army, and Hopkins had a demonstrated friendliness to veterans of the Southern cause. Professor of Greek Basil Gildersleeve, mainstay of the faculty, had been severely wounded while fighting with the rebels in the Shenandoah Valley. Sidney Lanier, lecturer in English, had served time in a prisoner-of-war camp after being captured on a blockade runner. Nevertheless, Miller's application caused considerable debate within the Board of Trustees. Ultimately the decision turned on an appeal to the personal beliefs of the founder of the university, the late Johns Hopkins himself. For Miller was black, and Hopkins had been a Quaker and an ardent abolitionist. Miller was admitted, the first African American graduate student of mathematics in the United States.

Miller's father, also named Kelly Miller, had been one of the small minority of free blacks in antebellum South Carolina. It is estimated that there were fewer than 10,000 such people in the state in 1860, compared to

400,000 enslaved blacks and 290,000 whites. But the elder Miller's "freedom" was sharply limited. In the decades before the crisis, whites in the South viewed free blacks with increasing suspicion. They were a challenge to the slave system by their very existence, and they excited alarm as potential fomenters of unrest among those still in full bondage. Laws were passed prohibiting manumission and restricting the fields of employment in which free blacks could work. Schemes were proposed to deport free blacks to Africa, or at least out of the slave states. When the war came, the elder Miller, a tenant farmer, was conscripted into the Confederate Army, essentially as a servant, a tactic employed to keep the population of free blacks under control during the rebellion.

The younger Kelly Miller was born in Winnsboro, in north-central South Carolina, in 1863. As of January 1 of that year, according to President Lincoln's Emancipation Proclamation, Miller's mother, because she resided within a state in rebellion against the United States, was "then, thenceforward, and forever free." In practice, since there was no one in her vicinity inclined to enforce this proclamation, she remained enslaved until 1865, when Union troops arrived in the town. The soldiers enjoyed playing with Kelly, an entertaining toddler, and suggested to his mother that they take him north, where they proposed to exhibit him as a sort of circus act. Later in life Miller would dryly describe this incident as "my first contact with Northern Culture." His mother refused to part with her son.

Miller's next encounter with Northerners was more positive. Even before the war was officially concluded, missionary societies, especially from the centers of abolitionist sentiment in New England, were dispatching hundreds of teachers to the South with the aim of bringing literacy to the formerly enslaved. Southern whites of the time disparaged them as "carpetbaggers," and later enlightened critics have accused these teachers of snooty condescension toward their "degraded" pupils. But the best of the missionary educators, mainly women, were struggling valiantly and sincerely with a challenge of enormous proportions. Well over 90% of adult blacks in the South were illiterate; hardly surprising when teaching them to read had been illegal. Miller later described the situation with characteristic wit: "While the slaveholder had proved beyond all possibility of doubt the incapacity of the Negro for knowledge, yet he, prudently enough, passed laws forbidding the attempt."

In 1860 there were approximately 20,000 African Americans residing in Miller's home county of Fairfield. Of these, he estimated that only 20 could read and write. The Northern teachers found, when they reached the South, that many of these illiterate adults had become fierce advocates of education for both themselves and their children. Miller's father sent all 10 of his children to school as soon as the opportunity presented itself. Miller always recalled with great affection the white women who first gave him a glimpse of learning. It was teachers such as these who brought the first inkling of mass education to the South, which had long resisted the idea for students of any color.

One of Miller's early teachers stood out in his recollection: Willard Richardson, a Presbyterian preacher who had studied at Hamilton College and Auburn Theological Seminary in New York State. He had had extensive experience as both a schoolteacher and a minister before serving as a chaplain for a New York regiment at the end of the Civil War. Richardson came to Winnsboro in 1869 as a missionary to the newly freed people of the region and established the still-existing Calvary Presbyterian Church, and the Fairfield Normal Institute. He had shepherded many students through the classical curriculum of Greek, Latin, and mathematics, and while he had no special affection for the last of these subjects, he was able to recognize that Kelly Miller's talent for it was beyond the usual. All of Miller's later career, in mathematics, sociology, and national political discourse, would ultimately flow from this missionary teacher's recognition. It was with Reverend Richardson's assistance that Miller gained admittance to Howard University in Washington, DC.

When Miller came to Howard in 1880, the institution was 13 years old. Its founding had emerged from a jumble of ideas proposed for managing the population of newly freed citizens. The social problems associated with this population were especially visible immediately after the war in the nation's capital, where many of those now designated as freedmen had migrated, with minimal resources. One leading idea was that religious instruction was crucial, and therefore resources should be devoted to training a cadre of black preachers. There were hopes that these preachers would be able to lead their flocks back to Africa, and then proceed to thoroughly Christianize that continent. Thus Howard University was first envisioned as a theological seminary. Others thought that the freedmen were most in need of basic education, not necessarily religious; hence Howard should include a general teacher training institute—a normal school, in the parlance of the time. But although the promulgation of Christianity remained important to the devout founders, and although teacher training would have a place, Howard University came to embody a more ambitious academic plan. It would not only aim to emulate the long-established liberal arts colleges of the country but would also aspire to a range of advanced professional education, including medicine and law, that would justify calling itself a university. These lofty goals were far from realized in the early years but helped to motivate several generations of students and faculty, one of whom was Kelly Miller.

Because the university was proposed to be in the federally administered District of Columbia, its establishment required an act of Congress. Those who had sought a seminary and a normal school had explicitly targeted a "colored" clientele, but the incorporation act that finally made its way to the desk of President Andrew Johnson made no mention of race, nor of sex, but only of "youth." The unstated implication that both women and African Americans were welcome was widely understood, though perhaps not by Johnson, whose endorsement of the university has puzzled historians, in view of his lack of sympathy for the freedmen. In fact, the first students to enroll were the daughters of white faculty members, but black students quickly became the majority, with coeducation firmly entrenched.

The university was named for one of its principal founders, General Oliver O. Howard, who had experienced classical collegiate training as a graduate of Bowdoin College, and professional training, as a graduate of West Point. He had seen extensive service and even lost an arm during the war. At its close, Howard had the gargantuan task of integrating the new citizens into the American economy, becoming the head of the Bureau of Refugees, Freedmen, and Abandoned Lands, known simply as the Freedmen's Bureau.

The early university existed on a meager diet of donations and what assistance General Howard could cobble together from the federal government. The severe national economic slump of 1873 pushed the university to the brink, but it did not fall. The Howard teachers in Miller's student days in the 1880s were all white with only one exception, and the course offerings in the collegiate department were largely the traditional classical curriculum, with a smattering of modern literature and physical sciences. Miller's mathematical background was solid, but he needed some help raising his Greek and Latin to acceptable standards. Thus he initially enrolled not in the college proper but in the Preparatory Department, where he spent two years. Such preparatory departments were a common feature of many colleges and universities of that time.

Miller treasured his training in the classical languages, although at times he seems to have felt that most of its value could have been imparted through translations. The attention given to Latin and Greek was controversial at Howard, as it was at other American colleges and universities. Some argued that the intricacies of these languages were especially effective in training the mind, or that the writings of Plato and Sophocles, of Virgil and Cicero, were of a depth and richness unequaled by more modern authors. More cynical observers felt that studying the "dead" languages was merely a way of signaling one's wealth and privilege. The question of teaching the classical curriculum was especially fraught for black students. From soon after emancipation there were many who felt that the education of the newly freed African Americans should emphasize the practical arts. Booker T. Washington most prominently promoted this idea. But others felt just as strongly that a focus on manual work would relegate black students to second-class status.

Miller was one of three graduates in the Howard class of 1886: two men and a woman. While he was studying at Howard, Miller supported himself by working for the Pension Office. This bureaucracy, with its need for clerks comfortable with numbers, was a direct result of the Civil War, as the

Kelly Miller, 1880–86. Courtesy of the Moorland-Spingarn Research Center, Howard University Archives, Howard University, Washington DC

federal government struggled with commitments to award pensions to veterans disabled by war service and to dependent survivors of these veterans. As a result of his government service Miller became acquainted with other technical specialists in government employ in Washington, in particular with the staff of the *Nautical Almanac*, headed by Simon Newcomb. Newcomb had been a mere computer when the *Nautical Almanac* was still in Cambridge, Massachusetts, but now had become one of the nation's foremost astronomers. In addition to his *Nautical Almanac* duties, Newcomb was commuting to Baltimore, part-time, as chairman of Johns Hopkins University's mathematics department, succeeding its founder, J. J. Sylvester (see chap. 8).

Miller was eager to pursue additional studies in the mathematical sciences and was especially curious about Hopkins. Newcomb could see Miller's talent, but feeling that the young man needed more preparation than

Howard University had provided, he recommended that Miller engage a tutor: Edgar Frisby, an English-born astronomer, who was one of Newcomb's lieutenants at the *Nautical Almanac*. Miller studied analytical geometry and integral calculus under Frisby for a year, after which Newcomb assisted Miller in applying to Hopkins. Newcomb was no egalitarian radical, but he had himself emerged from obscurity as a poor boy from rural Canada, and he knew how a word from a well-placed patron could influence a career.

At Hopkins, Miller took courses from Newcomb in astronomy, including celestial mechanics. He also attended classes in physics and in such mathematical subjects as the theory of numbers, the theory of functions, and quaternions. He took a class in mechanics from Christine Ladd-Franklin's husband, Fabian Franklin. The university of Miller's era was still experiencing the invigorating excitement of a new institution breaking new ground and was attracting a remarkable array of faculty and students who would go on to fame. During the two years that Miller attended, future US president Woodrow Wilson, who had just received his PhD at Hopkins, was lecturing on administration. Thomas Hunt Morgan, who would later receive the Nobel Prize for Physiology and Medicine, was a graduate student in biology. Frederick Jackson Turner, future renowned historian (see chap. 10), was a graduate student in history. It is unlikely that Miller had any contact with these individuals. Wilson, in particular, was a racial bigot. Miller had little contact even with his fellow students in mathematics, who treated him, he recalled later, "with cool, calculated civility."

Miller was forced to leave Hopkins in 1888 without a degree because of a sharp rise in tuition, stemming from the financial troubles of the Baltimore & Ohio Railroad. Stock in this company had been a central feature of Mr. Hopkins's founding bequest, but the instability of the railroad industry was a major economic reality of the late nineteenth century. Miller left Baltimore and returned to Washington to teach high school mathematics. He had aspired to be such a teacher, but he found that his graduate school training had spoiled him for dealing with beginners in geometry and algebra. Consequently, he jumped at an offer from Howard to teach more advanced mathematics. The number of students truly interested in the field was not sufficient to generate the excitement that he had experienced at Hopkins, but he was inspired to ponder deeply on mathematical pedagogy. He found there was a substantial group of students who required visual and tactile assistance to grasp mathematical concepts, and for them he devised pictures and three-dimensional models to aid their understanding. But for those who could perceive the abstract mathematical essence, as he himself had been able to do as a youth, these concrete methods were boring and unneeded. As he recalled from his own experience, "In my own case, when I was a boy in the primary grades, the teacher used to try to illustrate fractions by cutting an apple into halves, thirds, and fourths. I could grasp the meaning of those terms as readily in the abstract as in the concrete, and while slower minded pupils were trying to learn the meaning of fractions by this concrete method, I was more concerned with eating the apple." The difficulties of meeting the needs of both the gifted and the ordinary student of mathematics continued to perplex Miller, as it has many educators to the present.

Miller wrote a textbook on plane and solid geometry but could find no publisher willing to take a risk on a black author. No trace of his manuscript for this textbook has ever been found.

Miller soon found himself as the only black faculty member at Howard University, and for a time the only black mathematics professor in the United States, and he felt a keen sense of responsibility to represent his race. He had been conscious of race from an early age: sensitive to how "Negroes" (as he generally called African Americans) were perceived by whites and mindful of his own relatively privileged position. It was not long before he began to use his platform at Howard to engage in polemical discourse on racial issues.

The 1890s was a time when the last embers of the racial idealism that had lighted the Reconstruction Era were largely extinguished. Segregation policies were locked in place for decades by the Supreme Court in the 1896 *Plessy v. Ferguson* case, which ruled that it was lawful for a railroad to insist that white and black passengers use separate, allegedly equal, accommodations. In that same year there appeared an extensive analysis of "the negro problem" by Frederick Hoffman, a statistician working for the Prudential Insurance Company. Hoffman, breezily claiming lack of bias by virtue of his German birth, presented an extensive array of numerical data, all interpreted by him as showing the inherent inferiority of black Americans because of "race traits," impervious to environmental factors. Kelly Miller responded to Hoffman with a pamphlet in which he deftly demolished Hoffman's more absurd claims, especially the contention that the negro race in the United States was on a clear and certain path to extinction.

Having entered the realm of political discourse, Miller never left. His review of the Hoffman book was the first publication of the newly founded American Negro Academy, a group of black intellectuals that included the seventy-eight-year-old Alexander Crummell, a veteran of the abolition struggle, and the twenty-nine-year-old W. E. B. Du Bois, the first African American to earn a PhD (Harvard, 1895). Miller began to publish essays with regularity in such journals as the *Atlantic Monthly, The Dial, The Independent, Popular Science Monthly,* and *Educational Review.* He developed a supple and dignified prose style, spiced with flashes of wit. Racial issues were always central in Miller's writings, with the roles played by education and religion being frequent allied concerns. Sometimes he focused on individuals, writing essays on Thomas Jefferson, Theodore Roosevelt, and Walt Whitman, each one assessed in relation to African American life.

In 1905 Thomas Dixon Jr. quoted Miller as "the distinguished Negro teacher of Washington" in an article for the *Saturday Evening Post* titled "Booker T. Washington and the Negro." Washington by then was known nationwide as founder of Alabama's Tuskegee Institute, which embodied his philosophy of raising the status of black Americans through moral rigor and industrial training, while avoiding confrontation with white political power. Dixon claimed to find the aims of Washington's work "noble and inspiring" and to have been moved by Washington's personal history, recounted in *Up from Slavery* of 1901. Nevertheless, Dixon asserted that education was powerless to overcome the chasm between the Negro race and the "Anglo-Saxon." The only certain consequence of the efforts of

Washington and Miller would be to unfit the Negro for working under white command, with the result that the secret desire of "ninety-nine Negroes out of every hundred" would ruthlessly emerge: to "amalgamate" with the white race. This was an obsessive white fear dramatized in Dixon's novels, beginning with *The Leopard's Spots* of 1902, and most famously in *The Clansman* of 1905, which was adapted by filmmaker D. W. Griffith in 1915 as *The Birth of a Nation*, a landmark in cinematic technique. Dixon concluded that the only solution to the "Negro problem" was the old nineteenth-century fantasy of mass repatriation to Africa.

Miller responded to Dixon later in 1905 with a privately printed pamphlet, "As to the Leopard's Spots: An Open Letter to Thomas Dixon, Jr." Miller acknowledged the personal courtesy extended by Dixon but refused to countenance the wholesale vilification of the Negro race. He argued that there was no evidence that Negroes were inherently deficient in mental capacity and strenuously denied that he, or any large portion of blacks, favored amalgamation of the races. He did not deny the backwardness of his race, but he attributed their predicament primarily to environmental factors, which the passage of time would eventually mitigate.

By becoming vocal on race issues in the early twentieth century, Miller was inevitably drawn into the central ideological dispute in black America at the time, that between Booker T. Washington and W. E. B. Du Bois. Du Bois respected Washington's achievement at Tuskegee but felt that manual training of blacks should not be allowed to interfere with the nurturing of a black intellectual elite, "the talented tenth." Moreover, he saw Washington as too often meekly accommodating to white views, in return for being anointed by whites as the singular spokesman and designated patronage dispenser for all of black America. Du Bois first publicized his case against Washington in *The Souls of Black Folk* in 1903.

Miller tried to stake out an intermediate position between Washington and Du Bois, sometimes leaning one way and sometimes the other, so that he was known as a "straddler." Miller's own success in mathematics made him totally confident that intellectual talent was not racially specific, but the fact that he remained a unique example helped confirm his belief that he belonged to a race in need of uplift. This inclined him to hold that Washington's emphasis on industrial training was the most appropriate education for most black Americans. Yet he also believed that liberal arts college education, such as he had received himself, was necessary to provide leadership for the black community. At times Miller criticized both Washington's cautious politics and Du Bois's more radical stance. Miller declined to join the Niagara Movement, named for a 1905 gathering of Du Bois and other activists on the Canadian side of Niagara Falls. Yet Miller in 1910 was an early enrollee in the long-lived organization that was one of the principal legacies of the Niagara Movement, the National Association for the Advancement of Colored People (NAACP). Du Bois regretted Miller's vacillation, declaring after his death in 1939 that Miller had never harnessed his intellect to its full capability.

Miller believed that Christian faith and mathematically based scientific knowledge were the essential foundation for future advancement of African Americans. But although mathematics had provided Miller's entry into higher education, and into intellectual life generally, he came to find that its specialized study removed him too much from the social issues that increasingly occupied his attention. In 1900 he contributed a data-rich analysis of "The Education of the Negro" to the Annual Report of the US Commissioner of Education, and within a few years he was teaching courses in the new discipline of sociology at Howard. After 1907 he taught only sociology. Prior to his retirement in 1934 he also served for a time as dean of the College of Arts and Sciences. Before distancing himself from direct responsibility for mathematics he had set a high standard, paving the way for a strong commitment to the subject at Howard, which in the 1920s became the long-term home of the first two black Americans to earn a mathematics PhD: Elbert Cox (Cornell, 1925) and Dudley Woodard (University of Pennsylvania, 1928).

10

Frontiersmen
Herman Hollerith and E. H. Moore

A hole is thus punched corresponding to person, then a hole according as person is a male or female, another recording whether native or foreign born, another either white or colored, &c.

Herman Hollerith

CHICAGO, HAVING GROWN during the nineteenth century from a remote fur-trading post to become the railroad hub and meatpacking center of the Midwest, was chosen by the Congress of the United States to host a commemoration of the four hundredth anniversary of Christopher Columbus's famous voyage of 1492. This event, known officially as the World's Columbian Exposition, and less formally as the Chicago World's Fair, was open to the public between May and October 1893. Most of the fair's numerous visitors came to gawk at the exhibits of industrial innovation and to revel in the amusements, such as the gigantic revolving wheel designed by Mr. G. W. G. Ferris. But a small minority came for intellectual community. The organizers of the fair, following a model set by the 1889 World's Fair in Paris, arranged for a variety of religious, educational, and scholarly groups to hold meetings in conjunction with the fair. Thus it was that in the summer of 1893 historians and mathematicians gathered in separate conclaves in Chicago.

At the meeting of the American Historical Association, in July, Frederick Jackson Turner of the University of Wisconsin presented a paper claiming to explain the broad outlines of the country's development. This paper would become a landmark in the field, known by reputation, if not read, by every graduate student dipping a toe into American history. The data on

which Turner relied in his paper were significantly facilitated by a technological breakthrough that in the twentieth century would help launch the
computer industry, with far-reaching implications for mathematics and
beyond.

At the International Mathematical Congress, in August, Eliakim Hastings Moore, the acting head of the University of Chicago's mathematics
department, read a groundbreaking paper on a specialized topic that would
become a favorite of American mathematicians. Although the existence of
this particular paper is now little remarked upon except by a tiny contingent of historians of mathematics, it epitomized the emergence of pure
mathematics as the dominant concern of academic mathematicians in the
United States, a dominance that would persist virtually unchecked until
World War II. Moore himself would be the intellectual progenitor of a large
proportion of twentieth-century research mathematicians in the United
States.

Turner's paper was titled "The Significance of the Frontier in American History." He had been inspired by a remark of the superintendent of
the census, who had observed in 1891 that for the first time there was no
longer a clear "frontier line." Previously, the farthest extent of "settlement," meaning the regions of the country where people of European
heritage had established themselves, could be clearly distinguished from
"unsettled" areas by a line on the map. Now, however, the settlers had become so thoroughly dispersed as to make this impossible. The fact that the
superintendent could make this claim so soon after the 1890 census, and
that Turner could expound upon its implications at the 1893 Chicago
World's Fair, required significant technological innovation.

The Constitution of the United States, written in 1787, mandated a population census every ten years. Since the results were to be used to appor-
tion how many seats each state received in the House of Representatives,
the census was, and remains, charged with political significance. Moreover,
from the beginning of nationhood, the census required distinctions among
persons, specifically racial distinctions. The writers of the Constitution
stipulated, while carefully avoiding any form of the word "slave," that after
counting all "free persons" and indentured servants in a given state, but

excluding Indians, the total should be augmented by adding in "three-fifths of all other persons," a phrase that would reverberate through the decades. By the 1780s there was emerging a sharp divide in social structure and economic interests, South versus North, between those states wherein resided a large number of the aforesaid other persons, and those states where few lived. The three-fifths clause attempted to strike a balance of political power, promoted by James Madison, the Constitution's philosophical chieftain, slaveholder, and future president. The census was soon employed for probing a variety of non-constitutionally mandated distinctions as well. At no time has the census merely reported one number per state, as it might perhaps do, were one to interpret its mission in an extreme minimalist fashion.

As the country grew in the nineteenth century, each census required greater effort than the last, not merely to collect the data but also to compile it into useable form. The processing of the 1880 census was not completed until 1888. It had become a mind-numbingly boring, error-prone, clerical exercise of a magnitude rarely seen. Because the population was evidently continuing to grow at a rapid pace, those with sufficient imagination could foresee that processing the 1890 census would be gruesome indeed without some change in procedure.

On September 23, 1884, the US Patent Office recorded a submission from a twenty-four-year-old graduate of the Columbia School of Mines titled "Art of Compiling Statistics." By progressively improving the ideas of this initial submission, the young man would decisively win an 1889 competition conducted by the Bureau of the Census, which selected his system to mechanize the processing of the 1890 census. The inventor would go into business selling this technology, and the company he founded would, after he retired, become International Business Machines (IBM), which in the decades after World War II would dominate the computer industry and for a time would be the third-largest corporation in the world. The role of mathematics in the original census-tallying machinery was minimal. Conversely, the machinery had minimal influence on mathematics. But in the fullness of time, mathematics and the electronic computer, which in part grew out of the census machinery, would develop a relationship of considerable causal complexity, in both directions.

Herman Hollerith, the young inventor of 1884, had studied at Columbia with William Trowbridge, a West Point graduate (first in the class of 1848) who brought a combination of mathematical rigor and practical know-how to the teaching of mechanical engineering. It was through Trowbridge, a consultant at the Bureau of the Census, that Hollerith found employment at that agency and learned of the perplexities facing it. Even after establishing his entrepreneurial career, Hollerith retained sufficient connection with Columbia University that he was able to submit a dissertation on his census-tallying equipment to the School of Mines, which waived the residence requirement and awarded him a PhD in 1890.

Columbia had been founded as King's College in 1754, one of the nine colleges to emerge in the British North American colonies before the

Herman Hollerith. University Archives, Rare Book and Manuscript Library, Columbia University Libraries

Revolution. It grew substantially after the Civil War, and by the 1880s it was developing a core of ambitious young faculty members who would lay the groundwork for the major research university it would become in the twentieth century. In the mathematics department, Thomas Fiske, while still a graduate student in 1888, took the initiative to found a mathematical society like the one he had observed on a recent visit to England. This New York Mathematical Society began to hold regular meetings and soon attracted members beyond Columbia. The blossoming of this group into a national organization, the American Mathematical Society (AMS), was one of the early administrative triumphs of E. H. Moore.

Moore, born in 1862, had been privileged to receive the most rigorous mathematics education then available in the country, first at Woodward High School in Cincinnati, whose early days were glimpsed in chapter 4, and then at Yale, obtaining a bachelor's degree in 1883 and a PhD in 1885, both in mathematics. As noted in chapter 5, Moore was one of the rare students who impressed the great Yale mathematical physicist Willard Gibbs. Moore's dissertation, unlike that of Gibbs, was on an abstract topic: "Extensions of Certain Theorems of Clifford and Cayley in the Geometry of n Dimensions."

It was at Yale that Moore first met William Rainey Harper, then a classics professor and later to be the first president of the University of Chicago. In 1892, when Harper was assembling the first faculty at the new university, he was unable to snare a professor of established reputation to head the mathematics department, despite the largess of the founder, oil tycoon John D. Rockefeller. Harper's choice of Moore, as a stopgap, was abundantly rewarded. Moore built the Chicago mathematics department into the most productive in the country. Under his supervision it produced more than twice as many PhDs as any other institution in the United States. The students he personally supervised became leaders in the next generation of mathematical research and led productive departments at Harvard, Princeton, the University of Texas, and elsewhere. Moore also took an interest in school mathematics. His influence there was minimal, but he established a precedent, erratically observed in the twentieth century, for intervention in the schools by university mathematicians.

When Moore took charge at Chicago, he immediately strove to bring American mathematics abreast of the leading international developments in the field. When he learned of the proposal for a World's Congress in Mathematics and Astronomy at the 1893 World's Fair, he seized on it eagerly. He was especially keen to bring to the congress the man he considered the world's leading mathematician, Germany's Felix Klein. Moore not only secured Klein's attendance, but he also persuaded Klein to stay after the Chicago congress to give a series of lectures at Northwestern University, just north of the city. Furthermore, Moore succeeded in splitting the congress into two separate meetings: one for astronomy and one for mathematics. This symbolized a new age of autonomy for mathematics, no longer tied to physics and astronomy as it had been for the figures of earlier generations, such as Benjamin Peirce and Simon Newcomb.

This new mathematical independence was further ratified by E. H. Moore's own paper at the Mathematical Congress, titled "A Doubly-Infinite System of Simple Groups." This was a highly abstract exploration, far removed at the time from any scientific field. The general topic, the theory of groups, was of nineteenth-century origin. Much of the impetus had come from the problem of solving polynomial equations, such as

$$3x^5 + 7x^3 - 5x^2 + 2x - 7 = 0.$$

As with quadratic equations, familiar from school mathematics, both third- and fourth-degree equations can be solved by formulas involving the four usual arithmetic operations, plus extraction of roots. It was a great achievement of nineteenth-century mathematics to show that no such formula exists for solving fifth-degree equations (such as above) or higher. The key to this demonstration was to associate with any polynomial equation a structure called a group, consisting of a set of elements representing the symmetry properties of the equation. Any two of these elements can be combined to form another in the set. By the 1890s it was realized that groups come in an enormous number of varieties. Certain of these groups, designated as "simple," could be thought of as foundational for all the others, in rough analogy with the way all whole numbers are built from prime numbers. Moore's paper of 1893 was an early effort to identify some important families of simple groups. In the twentieth century, the theory of groups would become central to the major research area called abstract algebra, and the classification of simple groups would become a landmark project of the field. Much to the surprise of many, group theory also became immensely useful in physics.

In his 1893 research paper and subsequently, Moore showed himself to be fully committed to the new approaches to mathematics coming out of Europe, especially the language of "sets." This deceptively simple notion encompassed any collection of numbers or other objects. Beginning in the 1870s, the German mathematician Georg Cantor had produced astonishingly novel results by a close analysis of the process of counting the elements in a set. To count a finite set, one creates a one-to-one correspondence between the members of the set and a set of numbers, say, 1, 2, 3, 4, 5 if the set has five members. A fundamental fact is that a finite set cannot be put in one-to-one correspondence with a smaller subset. But this is not true for infinite sets. For example, the infinite set of numbers 1, 2, 3, 4, . . . can be put into one-to-one correspondence with the subset of even numbers 2, 4, 6, 8, . . . ($1\leftrightarrow2$, $2\leftrightarrow4$, $3\leftrightarrow6$, $4\leftrightarrow8$, . . .). No number of either set is omitted, and no number is paired twice. In this sense the two sets are of equal size. Using this definition of the size of a set, Cantor proceeded to show that there were different levels of infinity. In particular, the set of numbers 1, 2, 3, 4, . . . turns out to be smaller than the set of all numbers in a fixed interval. Between, say, 0 and 1 we have not only an infinity of fractions but also, less obviously, a more multitudinous infinity of numbers that cannot be written as ratios of whole numbers, such as $\pi/4$. Cantor's ideas, greatly elaborated, would become thoroughly embedded across mathematics in the twentieth century. But in the late nineteenth century there were still skeptics, such as Simon Newcomb, who insisted that it was "pure nonsense to talk about one infinity being greater or less than another." Moore engaged in correspondence with Newcomb in the late 1890s, attempting to convince him otherwise.

The counting problems of the census, concerned only with finite sets, were of a different kind. The solutions devised by Herman Hollerith involved a suite of mechanical and electrical devices, none of which in-

Mathematician E. H. Moore of the University of Chicago. Science Service Collection, National Museum of American History, Smithsonian Institution

volved any advanced mathematics, and none of which added to mathematical knowledge, initially. The first crucial innovation was to translate data on handwritten census tally sheets to patterns of holes punched in cards. Each individual was accorded a card, with the particular arrangement of holes on the card indicating the person's sex, race, occupation, or other information. The punching of the holes was a process that required development of special machinery to ensure accuracy and efficiency. Hollerith then devised a machine to "read" the card, by probing the card with pins, so that only where there was a hole would the pin pass through the card to make an electrical connection, resulting in advance of the appropriate counter. Thus, for example, if a card for a white, male, farmer passed through the machine, a counter for each of these categories would be increased by one. The card was made sturdy enough to allow passage through the card reading machine multiple times, for counting different categories or checking results.

Hollerith's card technology for recording and tabulating large sets of data was gradually perfected and widely adopted for a variety of purposes.

By the 1930s many businesses were using cards for record-keeping procedures such as payroll and inventory. Some data-intensive scientists, especially astronomers, were also finding the cards convenient. IBM had by then standardized an 80-column card and had developed keypunch machines that would change little for decades. Card processing became one leg of the mighty computer industry that blossomed after World War II. It served as a scaffolding for vastly more rapid and space-efficient electronic mechanisms that now dominate, with little evidence remaining of the old regime. Well into the 1980s cards remained a primary means of loading data and instructions into computers. As computer pioneer Grace Hopper (see chap. 15) recalled about her early career, "Back in those days, everybody was using punched cards, and they thought they'd use punched cards forever."

Kenneth Appel and Wolfgang Haken at the University of Illinois were still using cards in 1976 when they shook the confidence of some mathematicians' understanding of the nature of their endeavor by making significant use of computers in a theoretical mathematical demonstration. The problem they worked on had been unresolved for over 100 years: finding the minimum number of colors needed to color a map, so that no two bordering countries have the same color. It was long suspected that four colors would always suffice, but no demonstration had withstood close scrutiny. Appel and Haken were able to reduce the problem to checking a large, but finite, set of complicated map configurations. It was impractical to check these configurations by hand, but by employing over 1,000 hours of computer time they managed the task.

Some mathematicians were disquieted by Appel and Haken's resolution of the four-color problem, maintaining that a good mathematical proof should provide not merely a series of correct logical steps but an illumination of the reasons why a result is true. They found the Appel-Haken "proof" lacking in this insight. Although qualms about this particular application of computers to mathematics have subsequently declined, further inroads by computers into theoretical mathematics have periodically sparked renewed anxiety that the field might be in danger of losing its cherished distinctiveness, in which an incontrovertible proof is the emblem of fully understanding a knotty problem. The certainty that Abraham Lincoln had seen in mathematical demonstration (see chap. 3) does not seem well conveyed by a computer proof. Noncomputer proofs of extreme length have also contributed to doubts. The classification of finite simple groups, announced in the 1980s, initially approached 15,000 pages, impossible to grasp in totality. The 1990s debate on these matters was captured in an article in *Scientific American* magazine by journalist John Horgan, provocatively titled "The Death of Proof."

Kenneth Appel had received his PhD at the University of Michigan under the direction of Roger Lyndon. Lyndon in turn had been supervised at Harvard by Saunders Mac Lane. And Mac Lane had received crucial career advice at the University of Chicago from E. H. Moore, shortly before Moore died in 1932.

In the early decades of the twentieth century, at the same time that Hollerith's punched cards were becoming entrenched in data-processing machinery, mathematicians were building another foundation piece of the future computer age, an understanding of the scope and limitations of algorithms, the step-by-step procedures by which computations are accomplished. One of the major figures in this effort was the English mathematician Alan Turing, who in 1937 published a hugely influential paper on "Computable Numbers," in which he described what has become known as a Turing machine. This was an abstract symbol manipulator, which Turing visualized as an infinite tape divided into squares. When a square passes under a scanning head, a symbol can be written or erased on the square or the square can be left alone, depending on the state of the square and the scanning head. Remarkably, Turing was able to show that this device, though cumbersome, was universal. It could perform any symbol-manipulation algorithm, including any arithmetic computation. A more sophisticated machine, though it might be more efficient, could not perform any algorithm not performable by a Turing machine. Turing was also able to show that there are distinct limits to computability.

As he was on the verge of publishing this paper, Turing learned that some of his results had been duplicated by an American mathematician at Princeton University, Alonzo Church, though without Turing's helpful visual scheme. To learn more about Church's work, Turing went to Princeton for a year, and in fact received an American PhD under Church's direction. Church had received his own PhD at Princeton in the 1920s under Oswald Veblen, whose 1903 PhD was from the University of Chicago, under E. H. Moore.

Twenty-first-century mathematicians, even of the purest sort, increasingly use computers as experimental devices. Computer algebra systems are capable of quickly performing symbol manipulations that would take days or weeks by hand, computer graphics allow visualization of complicated geometric objects, and it is only with computers that the properties of enormously large numbers can be investigated. Announcements of a new largest prime number always involve computers. The progeny of Herman Hollerith and E. H. Moore have become intertwined in ways that neither could have foreseen.

11

Poetic Historian
E. T. Bell

1906

Bell's *Men of Mathematics* is poetic history—the kind that Homer wrote—
and the kind of history that many mathematicians want to read. It is often *not* the kind of history historians want to write.

John Ewing

S OON AFTER THE GREAT EARTHQUAKE struck San Francisco on the
morning of April 18, 1906, with fierce fires active or imminent
across the city, Eric Temple Bell, then twenty-three, sought to protect his
treasured copy of *Théorie des Nombres* by Édouard Lucas by burying it in
the yard of the boarding house in which he had been living. When he dug
it up a few days later he found the volume unusable for further study, but
he kept it as a memento of that vividly recalled period of his life. Bell would
use his experiences of 1906 as part of his science fiction novel *The Time
Stream*, serialized in 1931, by which time he was a professor of mathematics at the California Institute of Technology, in Pasadena. He would
also retain his interest in Lucas, championing the French mathematician's
work and publishing his own research papers inspired by it. The charred
copy of *Théorie des Nombres* has been preserved in the Caltech archives
since Bell's death in 1960.

Although Bell became prolific in both research mathematics and science fiction, it was a third literary genre, not foreshadowed in the 1906
quake, in which he would make his most substantial mark: popularization
of mathematics. He ranked the accomplishments of past mathematicians
and set standards of aspiration for mathematicians of the future. He effectively defined the contours of the mathematics profession for several

generations of mathematicians and nonmathematicians alike. To the exasperation of professional historians, Bell seduced many readers into ignoring blatant falsehoods and bigoted remarks. With a consistently lively prose style (his fiction writing pales in comparison) and a genius for expounding vivid anecdotes (supported by evidence or not), Bell romped with gusto through the decades and centuries, dispensing caustic commentary. In 1993 he became the subject of a biography himself, by another major mathematics popularizer, Constance Reid, who revealed aspects of Bell's early life that he had kept hidden even from his wife and only child.

Mathematics has been recognized as an esoteric subject since ancient times, with a few individuals evidently knowing far more, and caring far more, than the majority. Only with the rise of mass education did this imbalance come to be perceived as both a problem and an opportunity. The first half of the twentieth century saw a notable flourishing of attempts to make mathematical knowledge more accessible to a wider public, with books and articles emerging with regularity. In the United States in particular, rising levels of education made people more capable of reading such works, at the same time making them realize more fully how far above them the mathematicians were flying.

A special impetus was provided by the work of physicist Albert Einstein. In 1919 his general theory of relativity was dramatically confirmed by the observations conducted during a total solar eclipse, setting off massive publicity throughout the world. The combination of Einstein's charming personality, frizzy hair, and audacious ideas about space, time, and gravitation fascinated the general public. He was also widely admired (outside his native Germany) for his resolute resistance to the jingoism that had infected so many, including scientists, during the Great War of 1914–18. And the public perceived, vaguely but correctly, that Einstein's achievement owed something important to mathematics, thus helping to fuel a demand for popular accounts of the subject beyond school books.

The professionals were fascinated by Einstein's work as well. He had sought an appropriate framework for his innovative physical ideas, a flexible language in which to describe with precision the complex interaction of time, space, and matter. He found the tools he needed in nineteenth-century mathematics, specifically the non-Euclidean geometry of the German mathematician Bernhard Riemann and the tensor calculus of the Italians Gregorio Ricci and Tullio Levi-Civita. Many mathematicians were inspired to explore and further elaborate these fields. Bell remarked in 1922 that had he been 15 years younger he would have devoted himself to the mathematics of general relativity, but he was too committed to the theory of numbers, in which he had built up considerable specialized knowledge of no relevance to physics. Bell contented himself with teaching an occasional course on relativity and providing overviews of the theory in his popular mathematics writings.

His first such effort came early during the Einstein craze, in July 1920, when *Scientific American* magazine announced a contest to explain general relativity in no more than 3,000 nontechnical words. Bell, then a professor at the University of Washington, was among the 300 contestants who submitted an essay. He did not win the $5,000 prize, but his was one of only thirteen essays that were printed in their entirety in the book resulting from the contest. This was Bell's first effort at transmuting technical details into popular form, although it did not display much of the lively language that would become his trademark. He would go on to write nearly a dozen books explaining science and mathematics to the lay public, the last one finished just as he was dying in 1960.

═════

Bell had been born in Scotland in 1883 and emigrated in 1902 to California, where he obtained a bachelor's degree in mathematics from Stanford University in just two years. When the earthquake struck in 1906, Bell was teaching at a preparatory school in San Francisco. What biographer Reid discovered was that Bell's connection to Northern California was of longer standing than he pretended. He had in fact come over from Scotland with his family when he was only fifteen months old, living 12 years in San Jose. Bell was also secretive about his father's occupational history, which included the fish trade in Scotland and fruit growing in California. In neither place was the senior Bell blazingly successful, but there was enough money in the family that when he died in 1896 his son was able to return to England for private secondary schooling. Bell testified later that he was made into a mathematician by the influence of a teacher at the Bedford Modern School. Comparably rigorous schools would have been rare in California at the time.

After his stint in San Francisco, Bell would earn a master's degree from the University of Washington in 1908 and a PhD from New York City's Columbia University in 1912. He returned to the University of Washington as a professor from 1912 to 1926, before moving to a professorship at Caltech, where he remained to the end of his career. In moving among Caltech, Columbia, Stanford, and Washington, Bell received an excellent overview of the variety within the burgeoning American university system of the early twentieth century, and of the place of mathematics within this system.

Stanford, like the University of Chicago and Johns Hopkins University, was a post–Civil War private university, based on a great industrial fortune. It was only 10 years old when Bell attended, and was being buffeted by the intrusive meddling of the founder's widow, but it possessed a small core of stimulating mathematical scholars who inspired Bell. The University of Washington, a state university with origins in the region's territorial past, featured a Seattle campus situated amid great scenic beauty. It made few claims to academic distinction in Bell's day, but the head of the mathematics-astronomy department, R. E. Moritz, was building a fine mathematics library, much utilized by Bell. Moreover, Moritz was fascinated by mathematical history, culture, and gossip, as Bell would be fascinated in turn. Columbia, as noted in chapter 10, was a venerable institution, founded when New York was still a colony of Great Britain. By the time Bell was a graduate student, its gradual transformation into a modern research university was well launched. In its mathematics department Bell encountered professors whose interests would mark his later career: history (David Eugene Smith), research in number theory (Frank N. Cole), and expository writing about mathematics (Cassius J. Keyser). Finally, Bell joined the California Institute of Technology just as it was rapidly metamorphosing from its previous incarnation as a minor technical school. Under the presidency of physicist Robert Millikan (Nobel laureate, 1923) the former Throop Polytechnic Institute was becoming one of the leading centers of scientific research in the United States. Bell would here come in contact with students, professors, and guest speakers of the highest rank.

A college education was still a rare achievement for Americans in the 1920s and 1930s, but the old centers of learning on the East Coast no longer held a monopoly. The basis was being laid for a nationwide system of higher education, with varying levels of emphasis on teaching and research. In the second half of the twentieth century the PhD would become a prerequisite for almost all professors at universities and four-year colleges, even those claiming primarily a teaching mission. In Bell's day earning a PhD was less typical, and his research productivity at the University of Washington, with a heavy teaching load, was extraordinary.

As foreshadowed by the book Bell buried at the time of the San Francisco earthquake, his mathematical research was primarily in the area known as number theory, sometimes referred to as higher arithmetic. Mathematical research by the end of the nineteenth century was becoming organized into three main specialties: geometry, algebra, and analysis. The first two were vast generalizations of the school subjects going by those names, while the third area designated the elaboration of the differential and integral calculus pioneered in the seventeenth century. Number theory, focused on the properties of the ordinary counting numbers $(1, 2, 3, \ldots)$, could profitably be pursued with tools from all three of the above branches but was sometimes considered a branch of mathematics unto itself. Carl Friedrich Gauss, one of the giants of nineteenth-century mathematics, had famously been quoted as dubbing mathematics the "queen of the sciences" and number theory the "queen of mathematics." This was a formulation that Bell highly approved of and invoked often. Bell's first full-length popular mathematics book was *The Queen of the Sciences*, published in 1931 in conjunction with the Chicago's Century of Progress Exposition, the World's Fair of that year. Chapter VII of Bell's book was on number theory and titled "The Queen of Mathematics."

Queen of the Sciences surprised its publisher with its popularity among the general public. The mathematical professionals likewise gave it a warm reception; popular mathematics serves not only to entice those outside the field, but also to boost the morale of those already inside. The reviewer in the *American Mathematical Monthly*, official journal of the Mathematical Association of America, admitted to being "frankly enthusiastic" and professed a desire "to recommend the book to everybody. . . . The apprentice in the guild of mathematicians will be inspired. The university lecturer will put the book down feeling that he has reaffirmed his faith." The reviewer in the *Mathematics Teacher*, the official journal of the National Council of Teachers of Mathematics, opined that the book "should be read by every high school teacher." Both reviewers lauded Bell's ability to describe a wide range of mathematical ideas in concise, understandable language. Both reviewers also noted Bell's other career as a novelist; the true identity of "John Taine," the pseudonym under which Bell published most of his fiction, had recently been revealed. The *Mathematics Teacher* reviewer attributed the attractiveness of Bell's mathematical descriptions to his practice as a novelist.

Encouraged by this response, Bell began to devote a substantial part of his literary efforts to nonfiction. His next such book, published in 1933, was *Numerology*. This was an eccentric survey of humanity's tendency to attach mystical significance to striking numerical facts, from the ancient Babylonians, through medieval theologians, to contemporary scientists. Bell meandered from anecdotes about Hollywood (as a professor in nearby Pasadena, he claimed special knowledge) to chitchat about distinguished thinkers, from Plato to Einstein. The whole mishmash, including quick forays into subtle properties of numbers and seemingly serious commentary on the surprising utility of mathematics, was bathed in Bell's uninhibited style, disdainful of nonsense and flimflam.

In 1934 Bell produced another quirky book, *The Search for Truth*. It explored the subject of its ambitious title through a mixture of personal recollection (he describes an encounter with philosopher William James in San Francisco only days before the 1906 earthquake) and eclectic displays of erudition, not confined to science and mathematics. The biting comments of an unidentified personage named "Toby" are frequently quoted. Friends recognized Toby as the nickname of Bell's wife.

Bell next returned to a more sober product, *The Handmaiden of the Sciences*, which he cast as a companion to his earlier *Queen of the Sciences*. In the earlier book he had emphasized how mathematical concepts could be developed purely out of intellectual curiosity, irrespective of practical applications, while noting that some apparently useless notions had proved remarkably fruitful for the sciences. In the new book Bell reversed the emphasis, underlining the manifold ways in which mathematics had been explicitly developed as a tool of science, physics especially.

Handmaiden of the Sciences, as with Bell's earlier nonfiction, was published by the Baltimore firm of Williams and Wilkins. But even as he was working on that manuscript, he had landed a contract for a more ambitious book with the New York publisher Simon and Schuster. Both books would be published in 1937, with the Simon and Schuster volume becoming by far Bell's most popular title: *Men of Mathematics*.

Men of Mathematics consists of an introduction and twenty-eight biographical essays treating thirty-three individual mathematicians and one family of mathematicians (the Bernoullis). One chapter is devoted to three ancient Greeks. The nineteenth-century Russian Sonya Kovalevsky, the first woman to earn a doctoral degree in mathematics, shares a chapter with her German mentor, Karl Weierstrass. All the other chapters are focused on male European mathematicians who flourished between the seventeenth century and the early twentieth century. Amid deftly sketched mathematical ideas, Bell delights in colorful anecdotes and amusing foibles: "[Leibniz] was forever disentangling the genealogies of the semi-royal bastards whose descendants paid his generous wages, and proving with his unexcelled knowledge of the law their legitimate claims to the duchies into which their careless ancestors had neglected to fornicate them." He tosses off value judgments with flamboyant confidence. Throughout, mathematics is depicted as a supremely creative but human endeavor. As one reader remarked in 1945, "Since the appearance of Eric Temple Bell's *Men of Mathematics*, mathematicians have shown less fear of the idea of being counted as 'real persons.'"

Along with the humanizing, the book paints a strongly hierarchical picture of mathematical accomplishment. There is a clear suggestion that almost all the world's mathematical knowledge can be traced to those few individuals awarded a chapter heading, plus a smattering of others referenced in passing, chapters on some of whom Bell had to cut from the manuscript to comply with the publisher's word-count limit. Even within this small elite, distinctions needed to be made. The three greatest mathematicians of all, "in a class by themselves," Bell insisted, were Archimedes, Newton, and Gauss. This picture of the mathematics profession, exclusive yet human, although possibly discouraging to those Bell called "merely competent," proved hugely attractive to ambitious young novices. They dreamed that they too might grow up to be among those adding to the store of mathematical knowledge. "A club that you had to join," testified physicist Freeman Dyson. Despite Bell's overwhelming emphasis on male mathematicians, young women have not been immune to the book's charms. Julia Robinson, a notable late twentieth-century mathematician and sister of Bell's biographer Constance Reid, reported that it was *Men of Mathematics* that provided her first inkling of what it meant to be a mathematician.

Bell's style in *Men of Mathematics* was not so unrestrained as in *Numerology* or *Search for Truth*; Toby's periodic commentary does not appear, although the book is dedicated to her ("if she will have it"). But *Men of Mathematics* is by no means reserved in its judgments. Bell declares, with unmistakable homophobia, that no great mathematician has ever been a "pervert." Isaac Newton's appointment as England's Master of the Mint moves Bell to call out the "imbecility of the Anglo-Saxon breed." Blaise Pascal's contributions to French literature and religious thought are dismissed as "masochistic proclivities for self-torturing and profitless speculation." Jews are remarked for their "aggressive clannishness" and their "vicious academic hatred . . . when they disagree on purely scientific matters." When Taine Bell renewed the copyright for *Men of Mathematics* in 1965, five years after his father's death, these references to Jews were quietly replaced with more bland phrasing.

The book was generally well received by early readers, with some disclaimers. Bell was praised for being "witty and ironical," but his "stark,

Eric Temple Bell. Courtesy of the Archives, California Institute of Technology

frank humor" was sometimes felt to descend to "cheap jokes and silly remarks." Some declared him to be "superlatively illuminating," but others
warned against his propensity for exaggeration and his tendency "to emphasize various mathematical myths."

Bell would go on to write several more popular books that would elicit
similar responses. In the decades since his death, specialists in the history
of mathematics have become more outspoken and more specific in their
condemnation of Bell's historical misrepresentations, but mathematicians
have persistently shrugged off these criticisms. Reuben Hersh in 1990
lauded Bell's "narrative flair" in the "unrivaled collection of novelizations"
comprising *Men of Mathematics*. In 2006 Ian Stewart urged young mathematicians to "Read Eric Temple Bell's *Men of Mathematics*, still a cracking read even if some of the stories are invented and women are almost
invisible."

By the time Bell began writing his popular presentations on mathematics, being a mathematician had become a full-time profession, populated mainly by college and university professors plus some working in government or industry. Bell, with his multiple careers, was an outlier, and he

was fascinated by earlier part-time mathematicians. He delighted in the career of Pierre de Fermat (1601–65), calling him the "Prince of Amateurs" for making his contributions to mathematics on the side while serving as a judicial official. And Bell marveled at Gottfried Leibniz ("Master of All Trades"), who besides being the co-inventor of calculus with Isaac Newton in the seventeenth century was also a brilliant philosopher, historian, diplomat, and legal scholar. Twentieth-century mathematics was also a vastly more specialized activity than it had been in earlier eras, when an individual could bestride the entire field. In *Men of Mathematics* Bell acknowledged this change by declaring Henri Poincaré (1854–1912) to be "The Last Universalist."

Bell's own research was specialized indeed. One of number theory's attractions, often touted by its enthusiasts (including Bell), is that only a small amount of technical knowledge is needed to state and understand problems in the field. Yet the problems Bell tackled resist easy summary. Bell's biographer, Constance Reid, queried mathematicians for an assessment of the value of his number theory research and was especially impressed by Gian-Carlo Rota, an Italian American mathematician and philosopher. Rota, while having no personal knowledge of Bell, expressed qualified enthusiasm for a mysterious method called the "umbral calculus," which Bell had developed following suggestions in Lucas's *Théorie des Nombres*.

Much easier to understand than Bell's research are the statements of the classic number theory problems that he emphasizes in his popular books, notably the work of Fermat. The most celebrated problem in this line is the one generally referred to as Fermat's last theorem: the equation $x^n + y^n = z^n$ has no nonzero integer solutions if n is greater than 2. Fermat tantalizingly claimed, in a note he wrote in the margin of a book in 1637, to be able to demonstrate this fact, but that the margin was too small to contain the proof. For more than 350 years, mathematicians from all over the world strategized attacks on the problem, achieving gradually advancing partial successes. Bell's hero, Lucas, who had died in 1891, was rumored to have made an unpublished breakthrough, unhappily lost. The ultimate vanquishing of Fermat's last theorem, using methods that Fermat himself could not possibly have had in mind, was hailed by the aforementioned Gian-Carlo Rota as "a triumph of collaboration across frontiers and centuries. No domain of intellectual endeavor other than mathematics can claim such triumphs."

The mathematician given credit for finally pulling together all the threads of the Fermat problem was Andrew Wiles, who was educated in England but had been working in the United States for several years when he completed the proof in 1995. Wiles testified that he had been stalking Fermat since age ten, when he first read E. T. Bell's final book, *The Last Problem*.

12

Man of School Mathematics

Charles M. Austin

1914

Even at that time the progressive ideas of education were creeping into the schools. Principals and Superintendents were adopting the so-called progressive notions. They wanted to substitute easier subjects such as Social Science and Civics for Algebra and Geometry. Of course, we opposed this idea. These other subjects had no inherent value and they did not furnish a basis for future education.

Charles M. Austin

IN 1914, THE SAME YEAR that Europe descended into the Great War, later called World War I, a small group of mathematics teachers from high schools on the outskirts of Chicago began to meet regularly to help each other with educational challenges. They came to call themselves the Men's Mathematics Club of Chicago and Metropolitan Area, and their first chairman was Charles M. Austin, head of the mathematics department at the public high school in Oak Park.

As agitation for the United States to enter the war in Europe grew, a crisis closer to home was exercising the club members: threats to reduce or even eliminate required mathematics in the secondary schools. But Austin and his colleagues had mixed feelings when a national organization of college and university mathematics professors launched a study to shore up support of secondary school mathematics. Austin felt this effort should have been led by secondary schoolteachers themselves, if only they had had their own national organization. Austin therefore sought to create such a national group. The first meeting of the resulting National Council of

Teachers of Mathematics (NCTM) was held in 1920, with Austin elected as its first president.

With much of Europe recovering from the carnage of the war, the United States emerged as a global power in the 1920s and experienced vibrant economic growth. Mathematics in the schools remained embattled, however, despite the efforts of Austin and others. Not until the stimulus as a result of World War II in the 1940s and the Cold War in the 1950s would school mathematics once again be widely regarded as an indispensable subject. By the time of Charles Austin's death in 1967, a new set of controversies was engulfing school mathematics. His career illustrates the vulnerability of even the most privileged of the nation's high school mathematics teachers to an array of outside forces. The organizations that Austin spent much of his life nurturing have provided only modest refuge.

=====

Austin was born in 1874 in southwestern Ohio and began his teaching career in 1893, mere months after graduating from high school. This was standard procedure for many teachers of that era, but it was also clear to young Austin that to advance in the profession he would need further education. By attending Ohio Wesleyan between teaching stints, he was able to obtain a bachelor's degree in 1903. He became a high school mathematics teacher in 1904, still in rural Ohio. In 1912 he emerged on the bigger stage of suburban Chicago, landing the headship in mathematics at the Oak Park and River Forest Township High School. This was at a time when the population of high school students was doubling nationwide roughly every 10 years, and teachers with ambition could advance quickly.

The surge in students brought many challenges, beginning with the logistical and administrative problems associated with supervising a large quantity of immature human beings. The informal methods of the one-room schoolhouse were no longer adequate in the larger jurisdictions. Austin's early career saw the widespread adoption of educational management tools that have persisted ever since: segregation of students by age;

Oak Park and River Forest High School, Oak Park, Illinois, where Charles M. Austin taught for more than 30 years. Photo by Kmf164, taken on March 11, 2006. Licensed under the Creative Commons Attribution-Share Alike 2.5 generic license

strict classification of the curriculum into subject matter categories; monitoring of attendance; use of credit units to measure student progress; fixed daily schedules, with a bell marking the end of a period; and collection of graduation statistics. Questions of efficiency became paramount. In short, according to some alarmed observers then and now, the schools came to be organized as industrial operations, turning out a product. Others, such as Austin, largely accepted these developments as the inevitable price of progress.

Among the effects of this new regime in school education was a diminution in teacher autonomy and an increase in the number of administrators. Once upon a time (back in the nineteenth century), a school with more than one teacher had had a *principal teacher*. Now, larger schools simply had a *principal*, who did no teaching, while bossing those who did, assisted by a growing staff who likewise did no teaching. Districts with multiple schools were sprouting central offices, with layers of personnel yet further removed from the classroom. Conditions were ripe for workplace discord.

The burgeoning metropolis of Chicago proved to be especially susceptible. In the late nineteenth century, as the city became a center for transportation, meatpacking, and other industries, workers had organized in response to onerous conditions, with frequent conflict and unrest. Education, as another rapidly growing industry, experienced similar phenomena. The National Education Association (NEA) had been founded as early as 1857, but it would be many years before it represented the grievances of teachers. It was initially dominated by male college presidents, school superintendents, headmasters, and principals, even as the population of teachers became increasingly female. Much to the consternation of the male administrative elite, but entirely in keeping with the overall atmosphere in Chicago, many women teachers in the city and environs began to see analogies between their own situation and the oppressions visited on manual laborers. The teachers spoke out against the despotism of their male supervisors. Under the leadership of Margaret Haley the Chicago Teachers Federation became a labor union, fighting for improved working conditions, higher salaries, pensions, and tenure, and explicitly allying itself with the national labor activism of the American Federation of Labor under Samuel Gompers. In a triumphant moment of grassroots democracy, Haley helped engineer the election of Ella Flagg Young, Chicago's superintendent of schools, as the first woman president of the NEA in 1910.

This record of female teacher activism in Chicago suggests that the Men's Mathematics Club may have been regarded as a place of sanctuary by its members, and renders implausible the claim from 1965, in the official history of the club, that women teachers voluntarily formed a separate club because they objected to the men's cigar smoke. In fact, occasional proposals for more cooperation with female teachers evoked strong opposition within the Men's Club, with one club member in 1932 asserting that "the less we have to do with the women the better." It was not until 1972, five years after Charles Austin's death, that the two clubs merged into one, the Metropolitan Mathematics Club of Chicago.

The Men's Mathematics Club, as originally founded, was not especially interested in fighting administrative oppression but rather was concerned to preserve its members' relatively comfortable position in the educational hierarchy. Through much of the nineteenth century, mathematics had enjoyed an unquestioned place in the school curriculum. It was defended both for its practical utility and for its ability to train young minds in logical thinking, often referred to as the "mental discipline" thesis. But as the century came to an end, doubts were being raised. Ella Flagg Young herself had taught mathematics in her younger days but later grew disillusioned with the subject because it was "too mechanical." The rudiments of arithmetic were useful, but would most students need anything beyond that? In addition, psychologists were raising doubts about the alleged mind-training qualities of mathematics.

The sad fact was that the increase in the high school population meant that a larger proportion of students were in school under compulsion, either by parents or by attendance laws. Even students who enjoyed some aspects of school, such as Ernest Hemingway, a freshman at Charles Austin's Oak Park High School in 1913, dreaded mathematics. A smaller proportion of students were going to school for social polish, or for intellectual stimulation, and more with the aim of gaining employable skills. Most jobs seemed to demand little mathematics. Under these circumstances, there were calls for reducing mathematics requirements.

The early days of the Men's Mathematics Club of Chicago featured informal discussions over after-dinner cigars, enhancing group morale by swapping humorous stories about the tribulations of the classroom. But there gradually grew a determination to be more active, especially after the Mathematical Association of America (MAA) launched its secondary school curriculum study in 1916. The MAA was an offshoot of the American Mathematical Society (AMS), founded originally as the New York Mathematical Society in 1888. The AMS had come to represent the interests of college and university mathematical researchers as opposed to teachers. In 1915 those AMS members more interested in the teaching mission broke off to form the MAA. (Then and now there are many who belong to both organizations.) MAA members had a vested interest in school mathematics, inasmuch as the schools produced most of those who would later enter college and university mathematics classrooms.

The Men's Mathematics Club soon offered to assist the MAA study. The club formed a committee of six, Austin being one, to investigate "The Status of Mathematics in the Secondary Schools." The committee's method was to send a questionnaire to "prominent doctors, lawyers, merchants, bankers, etc., in the city of Chicago." These individuals were asked the following: whether their personal experience with high school mathematics had proved valuable; whether it could have been replaced with some other subject; whether they would insist that their children study algebra and geometry; whether algebra or geometry was of value in their business; and whether algebra and geometry should be retained in secondary schools.

The committee was gratified to find that the responders (who, though "prominent," were somehow at the same time declared to be "representative") overwhelmingly supported the importance of mathematics in secondary school, "some of them emphatically so." The committee published the statistics, along with some especially apt quotes. Some form of the mental discipline thesis was often invoked, as in this comment by a lawyer: "It is my judgement that the study of mathematics very materially assists in developing the reasoning faculties of the mind and powers of analysis."

Women were not entirely absent from consideration. There was at least one female respondent to the questionnaire, a social worker. (She had hated mathematics in school but later was glad to have been forced to take it "as general mental discipline.") The committee also noted that study of mathematics was important for girls as well as boys, because "the girls of today will be the mothers of tomorrow, and so will bear the greater share of responsibility in the education of the coming generation." This last remark seems to be a reference to some unstated assumptions about typical family dynamics, rather than an acknowledgment of the prevalence of female teachers (not all mothers, by any means).

After issuing the results of its survey, in January 1918, the Men's Mathematics Club committee proceeded to publish, in the three succeeding months, an article under the title "Valid Aims and Purposes for the Study of Mathematics in Secondary Schools." Here they offered an impassioned defense of their subject, citing authorities ancient and modern, from Plato to Alfred North Whitehead, not neglecting to mention Lincoln's encounter with Euclid (see chap. 3). Between January and February 1918 the lead author for the committee, Alfred Davis, changed his affiliation from being a secondary schoolteacher in Chicago to being a professor at the College of William and Mary in Virginia. This was emblematic of an issue that would bother Charles Austin to the end of his life: the difficulty for high school teachers to speak out independently of college and university teachers.

As noted, the Men's Mathematics Club became the core of the newly founded National Council of Teachers of Mathematics in 1920. Austin moved quickly to arrange an official journal for the NCTM by orchestrating the takeover of an existing journal, the *Mathematics Teacher*, which had been published since 1908 by a regional group, the Association of Teachers of Mathematics of the Middle States and Maryland. John R. Clark, an early member of the Men's Mathematics Club who had taught with Austin at Oak Park, was named editor. There was, however, no effort to make the NCTM a single-sex organization, like the club. Marie Gugle, assistant superintendent of schools in Columbus, Ohio, was elected to the first executive board and in 1926 became the fourth president.

The *Mathematics Teacher*, after it became an organ of the NCTM, was used as the primary vehicle for communicating to the community of mathematics teachers portions of the MAA's ongoing study of secondary school mathematics. The full report of that study did not appear until 1923: *The Reorganization of Mathematics in Secondary Education*. This document of more than 600 pages was hailed as landmark at the time, but in truth it had a modest influence on school mathematics, which remained under attack. The percentage of high school students taking algebra continued to fall.

The reorganization report highlights the precarious position of the high school mathematics teacher, even one as well placed as Charles Austin. The founding of the NCTM did not remove the many relative advantages enjoyed by college teachers compared to high school teachers: the former had more time to participate in activities outside the classroom and had more widely recognized rights to speak out on controversial subjects. Moreover, both college and high school mathematics teachers alike were finding new competitors claiming some jurisdiction over mathematics education.

Among such competitors were psychologists, who entered the field of mathematics education from more than one direction. Once in, they found plenty of fodder for their own research, independent of any felt need by teachers. Claims regarding mental discipline were fundamentally psychological, and indeed it was psychological research at the beginning of the twentieth century that had created much of the doubt that had grown up around the concept. This was so well recognized by the time of the reorganization report that its compilers devoted an entire chapter to "the present status of disciplinary values in education." Here was found another survey, this time of prominent psychologists. The conclusion reached was that there was support for a middle ground on the subject of mental discipline, in which some "transfer of training" did exist, depending on methods of teaching.

Another way in which psychology entered American school education in general, and mathematics education in particular, was through standardized testing. These tests emerged in the late nineteenth century and gained considerable popularity as a management tool during the early twentieth-century school population boom: for controlling entrance to and graduation from educational institutions; for measuring student progress in learning; for categorizing students by ability. This last aim, controversial then and now, was one that especially concerned Charles Austin. His Oak Park High School, finding it difficult to accommodate all students in the same elementary algebra classes, had sorted them into three groups: slow, medium, and fast. Austin and his colleagues had first tried relying on ratings provided by the lower schools from which the students had come. When this proved unsatisfactory, the Oak Park teachers experimented with alternatives, eventually settling on the Otis Intelligence Test, originally developed for the US Army during World War I. Austin in 1924 published data comparing Otis scores of "brightness" with scores in beginning algebra. He concluded that there was enough correlation to make it useful, although admitting that it did not account for student character flaws, such as laziness or indifference.

But it was the schools of education that were widely considered the most insidious competition for jurisdiction over mathematics classrooms. Normal schools, designed to train future teachers, had appeared before the Civil War, but by the twentieth century some teacher-training institutions were developing higher intellectual aspirations. Schools such as Teachers

College, at Columbia University, were not only training teachers but also cultivating theorists of education capable of appraising the field from a lofty elevation. Philosopher and psychologist John Dewey, from whom emanated much of what was referred to as "progressive education," was the presiding thinker at Teachers College for many years. Dewey had concluded that education needed to be totally reoriented to accommodate the industrializing twentieth-century world. Since students would be graduating into an ever-changing environment, general thinking and problem-solving skills were more valuable than any fixed bodies of knowledge. Many saw this as providing an intellectual justification for reducing mathematics in the schools.

Nevertheless, some of the faculty at the schools of education were considered friendly by schoolteachers such as Austin. Indeed, some of them were former teaching colleagues. John R. Clark and W. D. Reeve of Teachers College had both been schoolteachers in Chicago, where they had been early members of the Men's Mathematics Club. But other college educators were viewed as fundamentally hostile to mathematics. Most notorious was Dewey acolyte William Heard Kilpatrick, also of Teachers College, who despite studying with leading mathematicians at Johns Hopkins University and the University of Chicago had come to argue for diminishing the role of school mathematics, on the grounds that the subject was insufficiently connected with real life.

Especially among college-level mathematicians, schools of education were looked on as overly concerned with methods of teaching, at the expense of the mathematical subject matter. As early as 1932, Dartmouth College's John Wesley Young (chair of the effort to produce the reorganization report of 1923), publicly expressed fears that the schools of education were capturing the twelve-year-old NCTM and the schoolteachers it represented.

Charles Austin was one schoolteacher who was never captured by the mathematicians, the psychologists, or the educational theorists. He retired from Oak Park High School in 1945 at age seventy but kept current with developments in mathematics education. The 1964 annual meeting of the NCTM was held in Miami. There were 2,450 registered attendees, among them eighty-nine-year-old Austin, who had presided over the first such meeting, in Cleveland, in 1920, with 127 in attendance. At the Miami meeting banquet, Austin was seated at the head table with the current president and the board of directors. He was invited to make a few remarks. After surveying his surroundings, he said, "I look up and down this table, I see college people. Teachers, schoolteachers, founded this organization. Where are they on this board?"

13

Organization Man

E. B. Wilson

1922

E. B. was the only intelligent man I ever knew who liked committee meetings.

Paul A. Samuelson

IN 1922, EDWIN BIDWELL WILSON, chair of the physics department at the Massachusetts Institute of Technology, moved across the Charles River to Harvard's newly formed School of Public Health as a professor of vital statistics. In making this drastic career change Wilson was leaving physics just as that field was about to experience one of the most revolutionary periods in its entire history, and he was leaving MIT as that institution was in process of transforming from an engineering school to one of the preeminent centers of advanced scientific research in the world.

Wilson rarely alluded to missing these great developments. He pitched into health statistics at Harvard with the same vigor and resourcefulness that he had displayed in his previous work in pure mathematics, mathematical physics, and aeronautics. He would also make forays into economics and sociology, publishing articles on both these subjects. As a man who enjoyed displaying the breadth of his broad culture, he peppered his publications with references to Balzac, John Ruskin, George Bernard Shaw, Sophocles, and Winnie-the-Pooh. In a 1930 paper titled "Mathematics and Statistics," he treated *The Bridge of San Luis Rey*, Thornton Wilder's Pulitzer-Prize-winning novel of 1928, as an "interesting statistical treatise." Sometimes called a "Renaissance man," Wilson is remembered for no particular towering accomplishment, but he contributed to every field he touched. He was confident in his skills at navigating academic politics, certain of his command of technical details, and secure in the knowledge that he had apprenticed with one of the greatest scientists of the previous hundred years. Undergirding his self-assurance was a keen appreciation of the power of mathematics in all its luxuriant variety, pure and applied. But his judgment was not flawless. In the last decade of his life he chose to use his statistical skills on behalf of the tobacco industry. Subsequent developments have made this service seem craven, however sincere it may have been at the time.

<hr>

Wilson's mathematical talents were displayed early. His mother used to say that when he was four years old he learned the basic properties of arithmetic by folding and unfolding her 60-inch tape measure in all possible ways. He retained no memory of this, which would have been about 1883. His education continued rather unstructured until he went off to college. He had precociously wandered in and out of classes at the private secondary school headed by his father, in Middletown, Connecticut, without formal admission or being subjected to any school requirements. In this way he absorbed a classical education in Greek, Latin, and mathematics, standard for affluent young gentlemen of his generation. He also became well grounded in German. His father's school was a fitting school, in the parlance of the time. It aimed to prepare its students for the entrance examinations at one of the elite colleges of the Northeast, especially nearby Yale, his father's alma mater.

When he was fifteen, Wilson passed the Yale entrance exam, but his parents thought he was too young to go to college. The next year he did go, but to Harvard rather than Yale, because in the meantime his father had moved the family to Cambridge, Massachusetts. Harvard agreed to accept the Yale exam as evidence of young Wilson's fitness for admission.

Wilson took classes in both physics and mathematics at Harvard. Most significant for his later career was a course on quaternions from James Mill Peirce, the less brilliant brother of Charles Sanders Peirce (see chap. 8), both sons of Benjamin Peirce (see chaps. 1 and 6). By the time Wilson graduated from Harvard in 1899, age twenty, he was determined on a career in mathematics. The Harvard mathematicians advised him that, while he was welcome to remain there for a graduate degree, he would be better off going to Yale, where the opportunity for a subsequent faculty appointment would be greater. Thus Wilson came to know the great mathematical physicist J. Willard Gibbs (chap. 5), the supreme intellectual influence of his life.

At Harvard there had been only vague mutterings about Gibbs. A distant cousin of Benjamin's branch of the Peirce family, B. O. Peirce, from whom Wilson took courses in physics and differential geometry, once remarked that he thought Gibbs to be "a pretty able man." In view of B. O. Peirce's restrained personality, Wilson came to recognize this as effusive praise.

Wilson had written an honors thesis at Harvard on geometry, and therefore when he got to Yale he attached himself to the primary geometer on the staff, Percey Smith. Wilson's Yale dissertation, accepted for the PhD he earned in 1901, was written under Smith's direction soon after Wilson's arrival in New Haven: "The decomposition of the general collineation of space into three skew reflections."

Meanwhile, Wilson fell under the spell of Gibbs. To fill out his course schedule at Yale in his first year, Wilson was advised by the dean to take Gibbs's class in vector analysis. Wilson initially objected, believing, correctly, that this would duplicate much of his Harvard quaternion class. Wilson did take the class, found it easy, impressed Gibbs, and was impressed by Gibbs. There were two results: Wilson took every class offered by Gibbs from then on, and Wilson was invited by Gibbs to write up his lectures on vector analysis for publication in book form. Gibbs, almost totally preoccupied with his magisterial final book on statistical mechanics, gave minimal assistance to the twenty-two-year-old Wilson's shaping of the vector analysis material. The resulting volume stood for 50 years as the foundational treatment for efficient algebraic treatment of three-dimensional space for both physicists and mathematicians. It convinced many of the superiority of vector analysis to the related, but more cumbersome, quaternionic approach.

Another formative event while Wilson was at Yale was the visit of a star of French mathematics: Jacques Hadamard. Up to the advent of the Great War in 1914, Germany was the preferred destination for young American mathematicians seeking doctoral or postdoctoral study in Europe. But Hadamard convinced Wilson to deviate from the norm, maintaining that Paris had a higher concentration of first-rate mathematicians and mathematical physicists than any one German city.

Consequently, Wilson spent a postdoctoral stint in Paris, quickly picking up enough French-language facility to benefit from sitting in on courses at the École Normale Supérieure, the Collège de France, and the Sorbonne. As at his father's secondary school he was able to attend classes as often or as little as he pleased, without needing to satisfy any requirements. In this way he had the opportunity to hear many French luminaries, including the great Henri Poincaré, who gave surprisingly poor lectures, in contrast to his highly readable writings. Wilson found himself more impressed by the polished lectures of the less celebrated Joseph Boussinesq, a theoretical physicist of an older generation. Altogether, Wilson's French sojourn further expanded his already broad perspective on mathematics and physics.

Nevertheless, Wilson might have persisted with his original intention to specialize in geometry. But while he was in Paris, Gibbs suddenly died, age sixty-four, so that when Wilson returned to Yale, he not only received a faculty position in the mathematics department but was also asked to take on Gibbs's course work in mathematical physics. From this point on, Wilson never ceased to be eclectic. He began by writing variously on mathematical physics and pure mathematics. Even in areas where he made no claim to be a specialist, he enjoyed sticking his oar in by writing expository papers or book reviews.

Wilson's high competency in both mathematics and physics resulted in an offer of a position at MIT, which he accepted in 1907. It was here that he began to demonstrate skill with more practical research. MIT president Richard MacLaurin, a New Zealander educated as a mathematical physicist in Great Britain, returned from a vacation in England in 1912 alarmed about the likelihood of a war with Germany. MacLaurin proposed that MIT contribute to military preparedness by initiating a program in

Edwin Bidwell Wilson, ca. 1920. Emilio Segrè Visual Archives, American Institute of Physics

aeronautical engineering, and Wilson was tasked with teaching the relevant mathematics. A relationship between MIT and the US Navy was established, and by 1915 Wilson was writing reports for the government's National Advisory Committee for Aeronautics. In the 1950s this organization would become the National Aeronautics and Space Administration.

The pioneering nature of Wilson's aeronautical work in the 1910s is illuminated by noting that as late as 1906 the astronomer Simon Newcomb was still insisting that the problem of human aerial flight would require fundamental breakthroughs in basic physics. The era was awash in such breakthroughs, from relativity to radioactivity, but they were not needed to fly airplanes. In fact, as Wilson and his MIT colleagues showed, only a dozen years after the Wright brothers' initial success at Kitty Hawk, standard Newtonian mechanics and shrewd mathematical approximations could contribute to understanding eminently practical problems, such as flying an airplane faced with wind gusts. Wilson developed his aeronautical mathematics in constant interplay with measurements conducted on models in the MIT wind tunnel. A distinguished early graduate of the MIT aeronautics program was Donald W. Douglas, who went on to found the Douglas Airplane Company, maker of the DC-3 and other highly successful aircraft.

Wilson continued to publish on pure mathematics and theoretical physics even as he dove into aeronautics, but with less time to keep up with the latest developments. His administrative skills were also beginning to be recognized. He was named chair of the physics department while teaching mathematics and engineering courses. The apex of Wilson's high-wire juggling of MIT responsibilities came in 1920, when President MacLaurin unexpectedly died at the age of forty-nine. While a search was initiated for a new president, an administrative committee consisting of three faculty members was established to run the institution. Wilson, the junior member of the triumvirate, shouldered most of the work. He had to give up aeronautics, and his publishing slowed markedly. He did continue to deliver regular theoretical physics lectures in a remarkably comprehensive course on the constitution of matter. He managed this through judicious delegation, finding a bright, well-informed graduate student, Louise Eyre, who was able to help prepare the lectures. As he put it much later, "You can get a very remarkable job done by an able young person, if you give them a hard job to do." This was what Gibbs had done with him, and it was a technique he would repeatedly use himself in later years.

Beginning early in 1920, Wilson was de facto president of MIT for over two years. A new president was selected in mid-1920, but this man's disabling heart attack less than a month into his tenure kept Wilson's administrative committee in charge. But adept as he was at the work, Wilson had no wish to move permanently into academic administration. Moreover, as the months sped by, it became evident to Wilson that he would be in a delicate position whenever a new president finally did take command. With the history of executive decisions he was compiling, inevitably annoying some at MIT, he would have a hard time quietly settling back in as the chair of the physics department, or as a mere faculty member. Thus he was attracted by the prospect of a change of professional scene, without the requirement of a change of residence, which Harvard offered him with the vital statistics professorship in early 1922. And so Wilson effectively cut his ties with pure mathematics and mathematical physics and embarked on a new career. As one of his last acts in his old status he sent one of his students, R. Bruce Lindsay, over to Copenhagen to study with Niels Bohr, destined in the next few years to be the godfather of the revolutionary new theory of atomic structure: quantum mechanics. Wilson was not entirely happy with this new regime in physics, looking at it from afar. In 1927 he complained that "Physics consists of schizophrenic phantasy or manic ecstasy with a maximal obfuscation of complicated mathematical technique."

Wilson came to statistics with modest knowledge of the current state of the field, but with substantial background in the work of such early nineteenth-century giants as Gauss and Laplace. He would occasionally wield these celebrated names, along with that of Willard Gibbs (recalling that the latter's magnum opus was titled *Elementary Principles of Statistical Mechanics*), as rhetorical clubs with which to combat those he considered insufficiently grounded in foundational knowledge in statistics. At Harvard he now began to devote himself to the more recent technical literature, and to gaining familiarity with the problems that were uppermost at the School of Public Health. A theme of much of his writing on statistics, to the end of his life, was that no wise statistician applies a mathematical technique to analyze data without first having intimate knowledge of the context giving rise to the data.

It was not long before Wilson felt able to engage in debate about statistical matters related to biology. He was especially hard on Raymond Pearl, a biologist at Johns Hopkins, and even to some degree on Karl Pearson, a noted statistics innovator with whom Pearl had studied in England. Pearl and Pearson, according to Wilson, were artificially fitting curves to observational data by mathematical manipulation, without delving deeply enough into the mechanisms underlying the data. Wilson's animus led him to intervene in Pearl's career in 1929, upon learning that Harvard president A. Lawrence Lowell had offered Pearl a professorship in human biology. Wilson employed all his long-cultivated bureaucratic skills to manipulate the Harvard Board of Overseers into employing its rarely used power to overrule the president.

Wilson's statistical work was strongly collaborative. Vital statistics in the 1930s and 1940s proved to be more welcoming to women than many technical fields, and Wilson mentored several women who went on to substantial careers. With Ruth Puffer he wrote a comprehensive treatment of "Least Squares and Laws of Population Growth" (taking numerous swipes at Pearl). Puffer would go on to become the chief of the Department of Health Statistics of the Pan American Health Organization. With Jane Worcester, Wilson wrote 27 papers, notably on the statistical study of epidemics. She would become a nationally recognized expert on chronic diseases, achieve tenure at Harvard (one of the first women to do so), and eventually be named chair of the Department of Biostatistics (the successor of the Department of Vital Statistics, founded by Wilson).

Other statisticians quickly recognized Wilson's standing in the field. He was elected to be president of the American Statistical Association in 1929. This was but one of his many affiliations with scientific professional groups. He always held such groups in high regard, for disseminating ideas and fostering cooperation. The National Academy of Sciences was especially important to him. This organization, birthed during the Lincoln administration (see chap. 6), had been serving as an honorary society of the nation's best scientists. It was revitalized during World War I to carry out its original charter of providing scientific advice to the government. Wilson was elected to membership in 1919. Even before that, in 1915, he had been named managing editor of the academy's official journal, *Proceedings of the National Academy of Sciences*. He retained this position until his death in 1964.

In 1930, while continuing his position in the School of Public Health, Wilson was appointed to Harvard's newly founded department of sociology. He also began to teach a regular course in mathematical economics. These had been long-standing interests, dating back to his days at Yale in the early 1900s, when he had enjoyed conversing with economist Irving Fisher and sociologist William Graham Sumner. Fisher had the added attraction for Wilson of having been one of the best students of Willard Gibbs in the early 1890s. In 1912, while at MIT, Wilson had displayed his knowledge of Fisher, Gibbs, and Sumner in reviewing a book by the Italian economist Vilfredo Pareto. Also on display here was Wilson's conservatism. He complained that "the rising costs of government are bleeding the bourgeoisie for the support of the lower classes" and lamented attacks on "the formerly sacred institution of private property." He did not, however, transmit such views to his most celebrated economics student at Harvard, who opined that "It is paranoid to warn against inevitable slippery slopes . . . once individual commercial freedoms are in any way infringed upon." This was Paul Samuelson, 1970 Nobel laureate in economics, who expressed substantial indebtedness to Wilson.

After his retirement from Harvard in 1945, Wilson kept up a busy schedule of consulting, lecturing, and writing. He developed affiliations with the Office of Naval Research and the Council for Tobacco Research that continued to the end of his life. The latter organization was a creation of the tobacco industry in 1954, to respond to the growing evidence of a correlation between cigarette smoking and lung cancer deaths. But as every statistician, then and now, is at pains to caution, correlation does not imply causation. Wilson's scientific and political conservatism led him to ride this caution as hard as he possibly could.

The octogenarian Wilson, with undiminished sharpness and erudition, gave talks on "The Cigarette-Lung Cancer Enigma" into the 1960s. There was no convincing proof, he argued, that the "mild little cigarette" was causing cancer. It might be that there was a third, unrecognized, factor producing the correlation. This had been the case with malaria. It was not the "bad air" itself that caused the disease, but mosquitos that flourished in the conditions creating the bad air. Given the lack of clarity regarding a causal mechanism in the case of tobacco and cancer, and further given the evident pleasure that many obtained from tobacco, Wilson would not countenance a general call urging people to stop smoking cigarettes.

In private communications with colleagues on the Tobacco Council, Wilson engaged in ad hominem attacks on medical professionals who were taking an explicit anti-tobacco stance. Some he recalled as having been inferior students at Harvard; others he labeled as "publicity seekers." He admitted that the main desideratum of the tobacco industry was to find an alternative "whipping boy," which might divert attention from cigarettes as a cause of lung cancer. He was equally frank in admitting that no such alternative cause was anywhere on the horizon. The use of tobacco world-wide suggested to Wilson that there was "some substantial good to offset what everyone will admit are some of its bad effects when abused." The word or concept of "addiction" never appears in Wilson's commentary on tobacco.

In January 1964 the surgeon general of the United States released a re-port, *Smoking and Health*, emphatically asserting the ill effects of ciga-rette smoking. This report has come to be seen as a watershed moment in the debate about tobacco, but for Wilson, at the time, it was merely another set of data, complicating the picture. In August of that year, five months be-fore his death at eighty-five, Wilson published a two-page note titled "To-bacco Smoking and Longevity." This title, he noted, was identical to that of a paper published in 1938 by his old nemesis, the late Raymond Pearl. "As I do not find this reference in the recent report on *Smoking and Health*, I wish to make some comments upon it." Wilson proceeded to make some technical observations comparing Pearl's data with data provided in the surgeon general's report, noting at the end where he thought that further "careful study . . . should be made." Wilson was unmoved by emotional testimony being made by physicians describing the long-term ill effects of smoking. He refused to the end to deviate from his assumed role as a sta-tistical technician.

14

Versed in Math

Lillian R. Lieber and Hugh G. Lieber

1931

Most people, as you know,
regard mathematics as
the outstanding "pain in the neck"
of their school days.

Lillian R. Lieber

I N 1931, THE HUSBAND-AND-WIFE TEAM of Lillian and Hugh Lieber
published a short pamphlet titled *Non-Euclidean Geometry*, describing
the striking nineteenth-century discovery that the seemingly obvious
properties of parallel lines could be altered, while maintaining a logically
consistent geometric system. This was not a book for experts, but neither
was it appropriate for readers afraid of mathematical symbols. It proved to
be the first of a series of books published over the next thirty years, in
which the Liebers attempted to contribute to solving what they considered a pressing problem: the widespread public disdain for mathematics.

The presentation in all of these books was consistent and distinctive.
The text, written by Lillian, used short, left-justified lines, with abundant
white space. At least once this was described by a reviewer as "blank verse,"
but more often, and less ignorantly, as "free verse," for this was the era of
such poetic experimenters as e. e. cummings and Don Marquis. But Lillian
insisted from the beginning that she was simply trying to make her prose
more readable, without any versifying aspirations. Hugh's drawings,
although sometimes including geometric figures or algebraic symbols,
more often featured fantastic beings in a dream world ("Mathesis"), with
no attempt at realism.

142

After the 1960s, the whimsy and liberal social consciousness of the Lie-bers fell out of fashion as vehicles for the presentation of mathematics, even as E. T. Bell's cynical sniping remained popular (see chap. 11). The twenty-first century has seen a modest revival in interest in the Liebers' re-markable collaboration.

———

Lillian Lieber was born sometime in the late 1880s in Nikolayev, in what is now Ukraine. Although there was a tradition of learning in her family, had she remained in the highly patriarchal society into which she was born, her chances of attaining intellectual eminence would have been slim. Lillian's grandfather, Michel Bercinsky, had become a wealthy man in Pinsk. This town, like Nikolayev, was in the Pale of Settlement, the region in which the Jews of the Russian Empire were then tightly confined. Bercinsky was keen to give his sons the best education he could afford. But his daughters, in ac-cord with tradition, were deemed mere "empty nutshells" and given only rudimentary education.

One of Bercinsky's daughters, Haiye, by accidental circumstance be-came more educated than her father had intended. Her beauty attracted the attention of two men, both of whom recognized in her a thirst for knowl-edge, which they used in their courtship campaigns. The younger man, Haiye's true love, taught her algebra; the older man, a widower with four children, taught her Russian. Unhappily for Haiye, her father rejected the suit of the younger man because he had an aunt who had left the Jewish faith. Under protest, Haiye was married to the older suitor, Abraham Rosen-berg, a learned Hebrew scholar. She bore him four children but refused to take responsibility for the stepchildren, who were dispersed to other rela-tives. Haiye's youngest child, the only daughter, was the future Lillian Lie-ber. Before she was born, Rosenberg moved his family from Pinsk to Niko-layev, in part to put distance between Haiye and her former lover. Lillian's uncle, Nochim-Mayer Shaikevitsch, known by the pen name Shomer, was so fascinated by the history of this love triangle that he used it as the basis for one of his many novels.

In the 1880s, Jewish life under Czar Alexander III became increasingly restricted and subject to spasms of violence. A consequent wave of emigration sent many Russian Jews to the United States, including the Rosenberg family. Thus it was that in 1891 Abraham Rosenberg left his prominent place in the Jewish community of Nikolayev and entered a humbler existence in New York City, running a Hebrew printing shop. Lillian's cousin Miriam, Shomer's daughter, who had come to New York in 1889, described Abraham and Haiye (she would become known as Clara in the United States) as living a tragic existence in "an ugly little New York flat," with all that had been important to them in the old country "lopped off."

Being uprooted from the Old World, and coping with a new one, introduced confusion about ages and names. The passenger list for the *Fürst Bismarck*, the ship that carried the Rosenbergs to New York City in the summer of 1891, lists the daughter of the family as Helene Rosenberg, age three. But the US census of 1900 has her as Lillie Rosenberg, born in 1886, the date she would use for the rest of her life. By 1904 she was Lillian Rosanoff, a first-year student at Barnard College. Cousin Miriam asserted, with disapproval, that Lillian and two of her brothers had repudiated their parents by adopting the name Rosanoff, becoming thereby more Russian and less Jewish.

Whatever the trauma of being uprooted experienced by the parents, the educational opportunities for a girl were much greater in New York than they had been in the Pale of Settlement. The public schools in New York City in the late nineteenth century were riven by intense political fights and struggled to cope with huge increases in enrollment. But they charged no tuition and, all things considered, were welcoming to the ethnically diverse student body. A disciplined student, as Lillian evidently was, could gain substantial benefits from the opportunities offered.

It is not clear how she was able to afford Barnard College. Tuition plus associated expenses for one year would have been between $175 and $200, in the neighborhood of $5,000 in 2018 money. Possibly she was assisted by her brother Martin, about 12 years older. He was listed as sixteen-year-old Moses Rosenberg on the *Fürst Bismarck*, and again with that name in 1894 as a junior at the University of the City of New York (soon after renamed New York University). He graduated from that institution in 1895 as Martin Rosenberg, but by 1902 he had established a career in chemistry as Martin Rosanoff. The two siblings remained close until his death in 1951.

Cousin Miriam suggested, plausibly, that the change from Rosenberg to Rosanoff was a tactic to combat anti-Semitism. The educational institutions Lillian attended did indeed exhibit this affliction. At Columbia, situated amid the largest Jewish population in the country, the administration in the early 1900s was increasingly concerned that the Protestant elite, from "homes of refinement," were being scared away by the presence of too many Jews. Bryn Mawr, where Lillian was a research fellow from 1915 to 1917, had a clear record of turning away Jewish students and faculty.

Barnard, the first college for women in New York City, had been founded in 1889, after Columbia had rejected demands to become coeducational. By the time Lillian entered in 1904, Barnard was officially a constituent college of Columbia University, and it was located just across Broadway from the parent institution, on the Upper West Side of Manhattan. Some of the teachers taught only Barnard students, but other teachers taught at both Barnard and Columbia. This included a mathematics professor who would become important much later in Lillian's career: Edward Kasner. He was a specialist in geometry who would actively participate in the vibrant New York City mathematics scene for the next 50 years, a scene to which Lillian would make her own substantial contributions, beginning in the 1930s.

While Lillian was a student at Barnard, Franklin Delano Roosevelt was a student at the Columbia Law School. In the alphabetic listing of all the students of Columbia University for the academic year 1906-7, her name is directly under his. In stark contrast with Lillian's family, Roosevelt descended from wealthy forebears with deep roots in the United States, going back to colonial times. There is no reason to believe that Franklin and

Lillian ever met, then or later, but she would become an ardent supporter of his policies when he served as president of the United States. *The Einstein Theory of Relativity*, published by the Liebers in 1945, is dedicated to Roosevelt, "who saved the world from those forces of evil which sought to destroy Art and Science and the very Dignity of Man."

After graduating from Barnard, from 1908 to 1912, Lillian taught mathematics, first at the Normal College of New York (later called Hunter College) and then at a city high school. In the midst of this she found time to take graduate classes in physics and chemistry at Columbia, earning a master's degree in 1911.

In 1912 Lillian moved to Worcester, Massachusetts, becoming a doctoral student in chemistry at Clark University. Her advisor was her brother Martin, chair of the Clark chemistry department since 1907. She was awarded a PhD in 1914, with a dissertation titled "Theory of the Catalysis of Sugar Inversion by Acids" She spent another year teaching high school back in New York before being awarded the research fellowship at Bryn Mawr, outside Philadelphia, where she concentrated mainly on physics.

Thus by 1917 Lillian Rosanoff had accumulated an impressive set of academic credentials, but it would not translate into a permanent college-level position until the 1930s, when she attached herself to Long Island University, an institution with low prestige, no accreditation, and rickety finances. Her choices were surely constrained by anti-Semitism, misogyny, or both. After the Bryn Mawr fellowship she briefly taught physics at two small women's colleges, Wells College in Upstate New York, and the Connecticut College for Women, before returning to New York City to teach high school mathematics in Brooklyn. By 1923 her intellectual interests were becoming more focused on mathematics. She joined the Mathematical Association of America and began taking additional graduate courses in mathematics at Columbia. It was here that she met Hugh Lieber, who was in the process of earning a master's degree in mathematics and serving as a teaching assistant.

Hugh Gray Lieber was born in Missouri in 1896 but grew up in Oklahoma, obtaining a BA degree from the University of Oklahoma in 1919, after his studies were interrupted by stateside army service in World War I. He was not a Jew. His grandfather John Lieber had emigrated from Switzerland sometime before the Civil War. Hugh and Lillian married in 1926, and she would be known as Lillian R. Lieber for the remainder of her long life. Hugh continued as a graduate assistant at Columbia until 1928, without obtaining a further degree. In that year he joined the mathematics department at Long Island University in Brooklyn. LIU had been founded in 1926, specifically to be accessible to those, especially Jews, who were excluded from other colleges and universities in the region. It has been estimated that 75% of the initial class of 278 students were Jews. Then and later it attracted a large contingent of first-generation college students.

Also in the 1920s, Hugh began demonstrating, through a connection with the writer Theodore Dreiser, a deep interest in art, which would grow over the next decade and persist until his death in 1961. Dreiser, who in 1925 had scored his greatest literary success with the novel *An American Tragedy*, in 1926 published a book of poetry, *Moods: Cadenced and Declaimed*, in a limited edition of 550 signed copies. In 1928 Dreiser published a revision of this book, with illustrations by Hugh accompanying fifteen of the poems.

Lillian continued to teach high school mathematics until 1934, when she too joined the faculty at LIU. As noted, even before that year she and Hugh had begun their literary collaboration. They followed up *Non-Euclidean Geometry* of 1931 (a 34-page pamphlet) with *Galois and the Theory of Groups* in 1932. This 60-page pamphlet sketched the innovative ideas of French mathematician Évariste Galois, who had demonstrated that there exists no formula, analogous to the quadratic formula, for solving equations of the fifth degree or higher. Galois died in a duel in 1832, age twenty, before his importance was recognized, making him one of the most astonishing and romantic figures in the history of mathematics. Although the bare facts of his life seem dramatic enough, some biographers, notably E. T. Bell (see chap. 11), have been unable to resist insupportable embellishments. Lillian's brief account of Galois is relatively restrained but has not escaped criticism.

A short rave review of *Non-Euclidean Geometry* appeared in the *Mathematics Teacher*, authored by an LIU colleague, Adam J. Smith. A more critical review of both books, written by Heinrich W. Brinkmann, then at Harvard, was published in the *American Mathematical Monthly*. Brinkmann felt that Lillian placed misleading emphasis on certain technical details, especially in the Galois book, but he conceded that the difficulty of the task made it perhaps "the best that can be done in so few pages."

In 1934 Long Island University announced the formation of the Galois Institute of Mathematics and named Lillian Lieber as director. The institute invited research mathematicians from nearby colleges and universities to give talks, with undergraduates and high school students from across New York City encouraged to attend. The institute proceeded to publish the texts of many of these lectures and later sponsored the broadcast of some talks on the radio. Later, as the Galois Institute of Mathematics and Art, it would become the imprint for several of the Liebers' books.

The Galois Institute surely operated on a shoestring budget, with LIU in perilous financial condition during the Great Depression. Lillian was evidently harvesting good will in the local mathematics community that she had been cultivating for many years, possibly going back to her undergraduate days at Barnard during 1904–8. Among the speakers at the institute in the 1930s were Edward Kasner and C. J. Keyser from Columbia. Princeton, which in the 1930s was rising rapidly toward the pinnacle of American mathematics, was close enough to send speakers as well. Among the Princeton luminaries who spoke at the Galois Institute were Alonzo Church, Solomon Lefschetz, and Albert Tucker.

The Liebers published a pamphlet treatment of Einstein's special theory of relativity in 1936, contributing to the popularization boom surrounding that remarkable scientist, noted in chapter 11. Then in 1942 they produced the first edition of the book that would become their best known: *The Education of T. C. MITS: What Modern Mathematics Means to You*. This first edition was a spiral-bound softcover published by the Galois Institute, but the Liebers managed to interest a commercial publisher, W. W. Norton, in

Preface to *The Education of T. C. MITS*, drawings by Hugh Gray Lieber, words by Lillian R. Lieber, 1944. Reproduced with the permission of W. W. Norton

issuing an expanded hardcover in 1944, with blurbs from a gallery of highly distinguished commentators: C. J. Keyser of the Columbia mathematics department, novelist Dorothy Canfield Fisher, linguist S. I. Hayakawa, E. T. Bell (who had by this time achieved his own fame as a mathematics popularizer), and Albert Einstein. Moreover, a special US Armed Forces edition of the book was distributed to troops overseas during the last year of World War II, as part of the initiative of the Council on Books in Wartime, an organization in which W. W. Norton was a leader.

In *T. C. MITS* (the letters stand for *The Celebrated Man In The Street*) Lillian attempted to counter the common view of mathematics as a static endeavor with no scope for imagination or creativity. In making her case she ranged widely across the field. After surprising the unwary reader with the implications of simple numerical operations, such as repeatedly doubling the height of a stack of napkins, she wended her way into strange algebraic and geometric systems not far from the frontiers of mathematical research. She paused repeatedly to emphasize that mathematics was not a mere matter of manipulating numbers and symbols but was the great exemplar of the power of generalization and abstract thought: "a way of thinking, a way of life, VERY IMPORTANT FOR EVERYONE." She linked Mathematics (capitalized throughout) with Science and Art (also capitalized), all three revealing that "Internationalism and Democracy are very deep in the human spirit."

T. C. MITS was written at a time when the fate of the German Third Reich was still in question. Lillian's text decried Hitler's racial theories, while two of Hugh's drawings showed masses of people mesmerized by grotesque figures carrying the Nazi swastika.

In passing, Lillian generously described Columbia's Edward Kasner as "one of our great American mathematicians," while citing his introduction of the word "googol" for the number 1 followed by 100 zeros. This word, from which the name of the search engine and ubiquitous corporation derives, had first been mentioned by Kasner in a journal article in 1937, and then more widely disseminated in *Mathematics and the Imagination*, a successful popularization he coauthored in 1940 with his student James Newman, a lawyer and later journalist. Lillian repeated Kasner's claim that his young nephew made up the word "googol," an assertion that has been sometimes disputed and never confirmed.

In 1945 Hugh became head of the Department of Fine Arts at LIU, and Lillian replaced him as head of the Mathematics Department. Their next book was *Take a Number: Mathematics for the Two Billion*, published in 1946. This was an introduction to elementary algebra, in which the reader was asked to follow the guidance of a character called SAM, an embodiment of Lillian's admiring view of Science, Art, and Mathematics. Art, she explained, encompassed not merely visual art but also music and literature.

SAM would appear frequently in later books. He was present throughout *MITS, WITS, and Logic* of 1947, as a resource for the Man In The Street and the Woman In The Street as they navigated a world threatened by anti-

SAM-ites. This book eventually got around to covering standard topics in logic such as syllogisms and Boolean algebra but was frequently dominated by Lillian's impassioned social commentary. She made a plea for world peace and expressed alarm about the dangers of biological warfare and the atomic bomb, the latter used only two years earlier by the United States, to subdue Japan at the close of World War II. Lillian also stoutly defended abstraction in science and mathematics as well as in modern art. She noted that Hugh had coined a special word, "psyquaport," to describe his approach to portraiture, in which he aimed to include not only the psychological characteristics of the individual, but also the universal qualities that link all of humanity.

The Liebers were at the height of their fame in the late 1940s. In their unconventional way, Lillian and Hugh had made themselves, through the vehicle of mathematics, into an intellectual power couple in New York City. As recounted in the bestselling memoir of 1949, *Death Be Not Proud*, when journalist John Gunther sought to support the will to live of his gravely ill teenage son by nourishing the boy's curiosity about science and mathematics, the Liebers were called in. Young Johnny, battling a brain tumor, reported in a letter to a friend: "You may be interested to know that I was very lucky to meet Mr. and Mrs. Lieber who have written the books on higher Math in verse all full of wonderful illustrations. Mrs. L. is head Math Dept. and Mr. L. of Art at Long Island University." One day before he died, Johnny Gunther, expecting shortly to travel from New York City to spend the summer in rural Connecticut, packed a small collection of books. Among them were the Liebers' *Einstein Theory of Relativity* and *Galois and the Theory of Groups*.

Lillian and Hugh continued to turn out books into the early 1960s, mixing advanced technical concepts (lattice theory, transfinite arithmetic) with their strong social message. After Hugh's death in 1961, Lillian largely withdrew into obscurity, although she lived until 1986. A 723-page EdD dissertation on the history of Long Island University, completed in 1975 at Teachers College, Columbia University, makes no mention of Lillian or the Galois Institute. Hugh is listed among the faculty from 1927 to 1931, with no other mention.

Their books went out of print, but in the first decade of the twenty-first century a small publisher in Philadelphia, Paul Dry Books, arranged to republish *T. C. MITS*, *Einstein Theory of Relativity*, and *Infinity*, with new introductions by Lieber admirers. More recently, Maria Popova, a wide-ranging cultural commentator, has championed the Lieber books in her popular *Brain Pickings* blog. Although Lillian's political and social commentary and Hugh's drawings have never suited all readers, the Liebers' unique approach is still capable of creating new interest in mathematics.

15

Machine Whisperer

Grace Hopper

1941

I've been called an engineer, a programmer, systems analyst and everything under the sun but I still think my basic training is mathematics.

Grace Murray Hopper

O N DECEMBER 7, 1941, PhD mathematician Grace Murray Hopper was in the middle of a year of paid leave from Vassar College, taking advanced courses at New York University. Up to that time Hopper had appeared to be on an uncommon but not unprecedented career track, one of a small minority of female mathematics professors. But Japan's surprise attack on the US naval base at Pearl Harbor set in motion a sequence of events that would propel Hopper in a far more novel direction. She became a central player in a field that had not previously existed at all, for men or women: computer software development. Hopper lived long enough to see this field blossom into a vast worldwide enterprise largely distinct from academic mathematics, but she always valued her mathematical training as providing a crucial part of the foundation for her success in the new enterprise.

The elderly Hopper impressed the young computer scientist Jaron Lanier as "one no-nonsense, tough broad. No one was ever going to mess with her." But Hopper had not been so tough in the late 1940s and early 1950s, when the computer industry was struggling to be born and the success of her personal contribution was far from certain. Then Hopper had gone through periods of depression and alcohol dependence that nearly ended her career. She surmounted these difficulties to become, in her last decade, a confident and highly visible spokesperson for computer literacy, delighting audiences with television appearances on *60 Minutes* and *Late Night with David Letterman.* Hopper also became a symbol of female accomplishment in computer programming, a field that has nevertheless remained obstinately male dominated.

=====

She began life in comfortable circumstances in New York City as Grace Murray in 1906. Murray was educated at private schools and entered Vassar College as an undergraduate in 1924. This institution, in Poughkeepsie, New York, had been a source of scientifically educated young women since the days of Christine Ladd in the nineteenth century (see chap. 8). In Murray's day, Vassar mathematics was led by a distinguished mathematician of an earlier generation, Henry Seeley White, who had gotten his doctorate at Göttingen in Germany under the celebrated Felix Klein in 1891. White had been a great help in persuading Klein to make his inspiring visit to American in 1893 (see chap. 10). Murray valued White's teaching at Vassar, as well as that of Gertrude Smith, a 1901 Vassar graduate. Murray's wide-ranging curiosity led her to take or audit many subjects besides mathematics, a characteristic that would reappear throughout her life and would stand her in good stead. She graduated in 1928 with a bachelor of arts degree in mathematics and physics. Vassar awarded her a fellowship for graduate study in mathematics at Yale.

Although Yale would not admit undergraduate women until 1969, in 1928 it was more welcoming to women graduate students in mathematics than many universities of the time. Professor James Pierpont, a close contemporary of Henry White, had mentored several women to the PhD early in the century. And just prior to Murray's arrival, Pierpont had hired a younger man, Øystein Ore, who would also prove to be a source of support for female doctoral candidates. Ore had studied with the brilliant Emmy Noether in Germany, having followed the doctorate earned in his native Norway with a stint in Göttingen, before that world center of mathematics was shamefully torn apart by Hitler. Murray, who in 1930 became Grace Murray Hopper following her marriage to Vincent Foster Hopper, wrote a PhD dissertation in 1934 under Ore's direction. The topic was "irreducibility," bringing new insights to bear on the problem of determining whether an algebraic expression can be factored into simpler expressions.

While earning her doctorate at Yale, Hopper was employed as a teaching assistant in mathematics at Vassar. She remained in the Vassar mathematics department after earning her degree. There is no indication that she followed up on her dissertation or in any way tried to establish herself as a research mathematician. Her heavy teaching load, as a junior member of the faculty, would have made research difficult in any case. She also spent considerable time helping her husband with his doctoral research at Columbia University, on the history of European number symbolism up to the Renaissance. This resulted in one of her few publications in a mathematics journal, a short article on the mystical significance of the number 7 in ancient Greek writings. It was cited by her husband in the 1938 book growing out of his dissertation.

Hopper threw considerable energy into her teaching at Vassar. She enjoyed the challenge of juggling a variety of courses, from standards such as trigonometry and calculus, to less usual courses that no one else was inclined to teach, such as mechanical drawing, probability, and the calculus of finite differences. In many cases she was barely ahead of the class, teaching herself as she went along. She never liked to teach the same course in the same way twice, and she especially chafed against using conventional problems. Instead of filling and emptying cisterns with pipes of different capacities, as students had been doing for decades, Hopper had them analyzing traffic flow on a road system. In this way she built up a repertoire of skills and a confidence in her resourcefulness that would prove valuable when she moved into the computer field. Inspired by Walt Disney's pioneering animated movie of 1937, *Snow White and the Seven Dwarfs*, Hopper spiced up her mechanical drawing classes by supervising the student creation of a short movie illustrating, by animation, some properties of geometric curves. This too resulted in a short article by Hopper. Her students found her invigorating. At least two went on to PhDs in mathematics.

Hopper also continued to audit courses in the sciences that she had missed as an undergraduate, such as astronomy and geology. This made her sensitive to the special vocabularies and methodologies of the individual sciences, an understanding that would prove crucial in applying a general-purpose computer to these sciences.

But Hopper was restless, never quite content as a college teacher. It was this restlessness, and a desire to deepen her mathematical knowledge, that took her to New York University in the fall of 1941. Here she encountered another branch of the Göttingen tradition, in the person of Richard Courant, a Jewish mathematician who had left Germany early in the Hitler regime. At NYU, Courant established an institute partly modeled on Göttingen, but with greater focus on applied mathematics, in accordance with his own interests. With Courant, Hopper took rigorous courses in partial differential equations and the calculus of variations, basic to applying mathematics to the physical sciences.

It seems likely that Hopper would have eventually left college teaching in any case, but the sudden entry of the United States into World War II speeded up the process, and serendipitously landed her in the middle of the nascent computer industry. As with many Americans, the Pearl Harbor attack galvanized Hopper's patriotism. She sought vigorously to contribute to the war effort. There was a family history with the navy (one of her great-grandfathers had been an admiral), so she directed her efforts in that direction. For a thirty-five-year-old woman to actually enlist in the armed forces proved difficult, but she finally fought through the bureaucratic obstacles and joined the US Naval Reserve in December 1943. Some of the required classes at the Naval Reserve Midshipman's School for Women were a shock to her mathematical mind, as she was used to avoiding memorization by logically deriving consequences from a few fundamental principles. This procedure did not work for the defiantly illogical organization chart of the US Navy. Hopper did, however, enjoy having mundane decisions taken out of her control: "I didn't even have to figure out what I was going to cook for dinner."

Hopper was commissioned as a lieutenant (junior grade) in the summer of 1944, and she was almost immediately assigned to work on a project at

Harvard University supported by the navy. The military, famous for making questionable duty assignments, in this case hit the mark. The Harvard project was being run by Howard Aiken, who in the late 1930s had become interested in advancing the technology of calculating machines to compute approximate solutions to differential equations of significance to physics. Solving a differential equation essentially means finding a curve, a surface, or a higher-dimensional analogue, satisfying certain conditions. For example, one may wish to find the trajectory of a projectile, given assumptions on all the forces acting on that projectile. Only rarely is it possible to find an explicit formula for the desired curve or surface, in effect defining it for an infinity of points. In the absence of a formula, one attempts to approximate the desired solution by a finite table of numerical values. A basic tool for such work is the calculus of finite differences, methods of which Hopper had first encountered as a teacher at Vassar and then again in more sophisticated form under Courant at NYU.

Aiken established a relationship with International Business Machines (IBM) to construct the machine he envisioned, the Automatic Sequence Controlled Calculator. When World War II came, he went into the navy, with the IBM machine, later dubbed Mark I, still under construction. Shortly before Hopper's arrival, the Mark I had at last been installed at Harvard, with Aiken assigned in charge by the navy to use it to make calculations, first for the Bureau of Ships and then for the Bureau of Ordnance. It would also be appropriated for some work related to the Manhattan Project, the crash program to develop the atomic bomb.

Some of the central features of the Mark I were already obsolete. For example, it performed arithmetic in base 10, whereas base 2 (0-1, true-false, yes-no, on-off) would soon be generally recognized as superior. And although Mark I was powered by electricity, it still relied on many physically moving parts click-clacking away, placing severe constraints on operational speed. Technology to take fuller advantage of the near-light-speed transmission of electricity was in fact already available, as evidenced by the Electronic Numerical Integrator and Computer (ENIAC) being constructed contemporaneously at the University of Pennsylvania. ENIAC was fully electronic, using vacuum tubes as switching devices. Nevertheless, the

Lt. Grace Hopper at Harvard, 1944–46. Grace Murray Hopper Collection, Archives Center, National Museum of American History, Smithsonian Institution

Mark I's capability for performing a lengthy numerical computation, without human intervention after the initial data entry, was a breakthrough. Preparing a calculation involved feeding in a length of paper tape on which holes had been strategically punched. This helped the Mark I's operators, notably Grace Hopper, to conceive of "programming" as separate from the machine itself.

In the long run the most significant influence of the Mark I was as a training ground for the people involved. Many Mark I alumni, Grace Hopper being only one, became significant contributors to the computer industry in the 1950s and beyond. Tackling the intricacies of 1940s computing inspired them to imagine far greater possibilities in the years ahead. Surprisingly, Howard Aiken was at first not convinced of that bright future. In 1948 he infamously opined that the United States would never need more than five or six computers.

Hopper left active duty in 1946, while remaining in the naval reserve. Now divorced from her husband, she elected not to return to Vassar, instead staying on as a research fellow at Harvard, where she worked on the Mark II and Mark III, successors to the Mark I. In 1949 she moved to Philadelphia to join the company founded by J. Presper Eckert and John Mauchly, the designers of the ENIAC. The enterprise was in perilous financial condition, and it was at this time that Hopper's drinking problem became most acute. With the assistance of close colleagues, she managed to recover from alcohol dependency. Meanwhile, her job was saved when the Eckert-Mauchly Computer Corporation was bought by Remington Rand, which in 1955 became part of Sperry Rand. Hopper stayed with Sperry Rand until her retirement in 1971.

Eckert and Mauchly, in building the ENIAC and starting on a successor called the Electronic Discrete Variable Computer (EDVAC), had come to appreciate the importance and the difficulty of the "setup," which came to be called "programming" the computer. A bad setup could negate the capabilities of the machine, no matter how spectacular the technology. Spurred on by the extraordinary genius of the Hungarian émigré mathematician John von Neumann, a pervasive influence on almost all computing projects of the era, Eckert and Mauchly had further come to see that the data and the setup instructions could both be stored in the machine. When they began building computers commercially, this "stored program" concept became a central design feature.

From her Harvard experiences, Hopper was in full agreement with regard to the importance of the stored program. At Harvard she and Aiken had also observed the crucial fact that many numerical computations require repeated use of the same basic steps, perhaps with differing inputs. Out of this recognition was conceived the idea of what came to be called a subroutine within a larger program. Indeed, the same subroutine might be useful in many different programs. A library of subroutines could be established, permanently stored in some fashion, with particular routines called upon when needed. Hopper applied these ideas to Eckert and Mauchly's daunting project to create a commercially viable general-purpose computer, which they called the Universal Automatic Computer, or UNIVAC.

She put together a "program-making" program to automatically "compile" appropriate subroutines into a larger whole to accomplish a specific purpose. Within a few years Hopper and other computer professionals were giving the term "compiler" the meaning it has today: a program to translate instructions in a "high-level" language, readable by people, into a form executable by a computer.

In the 1950s, as in the late nineteenth century, when Herman Hollerith developed his punched-card tabulators (see chap. 10), the US Census Bureau was in special need of assistance in handling masses of data. The first UNIVAC was successfully delivered to the Census Bureau in 1951 and proved a great success. By 1954 there were 20 machines in operation in a variety of government and business environments. Although some users were still solving differential equations and computing mathematical tables, others were applying the machine to tasks such as payroll, inventory, and billing. In consequence, the UNIVAC came to be seen not just as a giant calculator but as a general-purpose information-processing device, with far-reaching implications for the future of the technology.

Hopper's emphasis on automating the writing of computer programs proved an enduring goal for the development of what came to be called computer "software." Her overriding aim, as she later put it, was "to make it easier for people to use computers." She wanted to make computer commands easier to understand, without requiring deep knowledge of the inner workings of the machine. By 1955 she had created FLOW-MATIC, the first English-language data-processing language. FLOW-MATIC was short-lived and specific to UNIVAC machines but was a major influence on the long-lived computer language COBOL (short for COmmon Business Oriented Language), which became portable across the machines of multiple manufacturers. COBOL was the product of a committee convened in 1959 by the US government, and its popularity was ensured when the government announced that it would buy or lease only computers able to run COBOL. Hopper was a member of the executive oversight committee for the language.

FORTRAN (FORmula TRANslator), another long-lived computer language that emerged shortly before COBOL, was only slightly more verbose than ordinary algebra. With COBOL, in contrast, following Hopper's practice in FLOW-MATIC, algebraic operations were written out in words, and variables could be assigned long names. The hope, sometimes illusory, was that COBOL programs would be readily understandable not only to the original programmer but to an untutored computer user.

Before retiring from Sperry Rand in 1971, Hopper began doing some part-time teaching at the University of Pennsylvania. She continued teaching in the 1970s, at George Washington University as well. Meanwhile her COBOL expertise resulted in extension of her navy service. Thus, although she was involuntarily retired in 1966, she was recalled the next year, first to head up COBOL standardization in the navy, and later to lead navy computer programming efforts in general. The navy regularly promoted her, and she reached the rank of rear admiral before finally being ceremoniously retired in 1986 aboard the nineteenth-century warship the USS *Constitution* ("Old Ironsides").

Hopper's name has continued to resonate after her death. In 2017, Yale University, after much debate, elected to remove the name of John C. Calhoun from one of its undergraduate residential colleges. Calhoun (Yale class of 1804) was one of the most prominent American political figures of the first half of the nineteenth century. He served as congressman, senator, secretary of war, secretary of state, and vice president. But Calhoun was also a leading proponent of white supremacy and race-based slavery. Yale changed the name of the college to honor Grace Murray Hopper.

16

Survivor

Izaak Wirszup

Poland was, so to speak, a new country, and in academic literature there were not enough Polish books in higher mathematics, so we had to study from German books and from Russian books and from French books too. There were some Polish books, but you couldn't exist with one language.

Izaak Wirszup

IN APRIL 1956 A RECENT PhD recipient, employed by the University of Chicago, was scheduled to speak at a meeting of the National Council of Teachers of Mathematics (NCTM) in Milwaukee, Wisconsin. The speaker did not propose to expound on the theory of infinite series, the subject of his dissertation, nor on any topic related to the world-renowned mathematical research of his dissertation advisor, the University of Chicago mathematician Antoni Zygmund. Nor was the speaker an untried novice in education or in any other respect. He was Izaak Wirszup, a forty-one-year-old Holocaust survivor, an émigré who had first studied with Zygmund in Poland before World War II. In Milwaukee, Wirszup was planning to speak on enriching the school curriculum by "broadening and vivifying the student's significant mathematical experience." He proposed illuminating the subject with his knowledge of pedagogical thought in Eastern Europe. And because Wirszup intended to make favorable remarks about aspects of the Soviet Union's education system, in the midst of the Cold War, his University of Chicago colleagues feared heckling and even physical violence. They therefore sent two of their number to Milwaukee with Wirszup to act as bodyguards.

As it turned out, the reception of Wirszup's talk was polite and sufficiently positive to confirm him in the decision he had already taken to devote the rest of his career not to abstract mathematical research like Zygmund, but to improving school mathematics. Less than two years later, in
October 1957, the Soviet Union launched Sputnik, the first artificial satellite to orbit the earth, and Wirszup's views on the vitality of Soviet education became abruptly more convincing.

Wirszup had no illusions about international politics. He grew up in a city
with a long history of disputed ownership, known today as Vilnius, Lithuania. From the late eighteenth century to the early twentieth, save for a
brief respite under Napoleon, the city had been an often-discontented
part of the Russian Empire. In 1915, the year of Wirszup's birth, it had
been occupied by the German Army as it pushed east in World War I. Following violent strife after the war, it became Wilno, in the extreme northeast arm of the newly independent nation of Poland. There had been no
such nation since the eighteenth century, when Polish-speaking peoples
had been partitioned among Austria, Prussia, and Russia. Even before the
1918 armistice, Polish nationalists were moving quickly to achieve their
long-deferred dreams, taking advantage of world opinion in their favor.
US president Woodrow Wilson had listed Polish independence as Point
Thirteen of his much-publicized plan for the postwar world, known as the
Fourteen Points.

Growing up in such a contested region, it was natural for Wirszup to
learn multiple languages, and this was especially necessary for someone
with intellectual aspirations. When he began studying mathematics at the
university, Wirszup came to realize that much of mathematical literature
was written in French and German, and some important work in English
and Russian as well. But his first two languages, for home and early schooling, were Polish and Yiddish. Wilno was a major Jewish center in Wirszup's
youth, with Jews constituting as much as 40% of the city's population.
Wirszup's father was a successful candy and chocolate manufacturer, able

to provide his children the best education. He encouraged Izaak in a scholarly direction, even though academic positions for Jews were limited throughout Poland. Discrimination against minorities in Poland was widespread in the 1930s. Jews, Lithuanians, Russians, and Ukrainians all endured low status at the University of Wilno, known then as the University Stefan Batory, after the famed Polish king who founded it in the 1500s. Many professors banned Jews from their classrooms. When Wirszup entered, at age seventeen, he was fortunate that one professor who rejected this policy was Antoni Zygmund, a Polish Catholic from humble circumstances in Warsaw.

By the early 1930s, Zygmund (born in 1900) had already established an international reputation. He had spent time at both Oxford and Cambridge and had collaborated with leading British and American mathematicians. His specialty was trigonometric series, a subject rich in applications to physics, while also giving rise to many subtle theoretical problems. When Wirszup first met Zygmund, the latter was putting together his book *Trigonometrical Series*, a definitive tome on the subject as of its publication in 1935. Early in the nineteenth century, the French mathematician Joseph Fourier had demonstrated that undulating sine and cosine curves were useful for more than describing periodic phenomena. By combining multiple sine and cosine curves of varying amplitudes and periods, it was possible to approximately represent even very irregular curves. The resulting "Fourier series" have ever since been a crucial tool in solving partial differential equations, a pervasive project in mathematical physics, wherever one quantity depends on two or more other quantities.

Zygmund was not only a brilliant mathematician in his own right, he was also excellent at judging mathematical talent in others, and nurturing it when he found it. At Wilno his great discovery was Józef Marcinkiewicz, five years older than Wirszup. Wirszup was a good student, but Marcinkiewicz was something far rarer—an original mathematical thinker, even as an undergraduate—a "genius type," as Wirszup remembered him. Marcinkiewicz and Zygmund soon embarked on a fruitful collaboration, heartbreakingly cut short by the events of 1939.

In the late 1930s, Poland found itself squeezed between two brutal dictatorships, both of which coveted Polish lands: Adolf Hitler's Germany on the west and Joseph Stalin's Soviet Union on the east. In 1939 the two murderous regimes signed a nonaggression pact and came to an agreement on the repartitioning of Poland. In September of that year they both invaded, beginning the European phase of World War II. Within weeks, Poland once again ceased to exist. Lithuania, retaining nominal sovereignty for a time, was allocated to the Soviet sphere, with the city it called Vilnius as part of its territory.

Wirszup had finished a master's degree just before the outbreak of the war and had started to work on a doctorate, but when the Soviets took control, the university was closed. Wirszup got a job at the State Technical Institute, teaching analytic geometry and calculus to engineering students. Zygmund and Marcinkiewicz, meanwhile, had both been mobilized into the Polish Army during the buildup to the war. Zygmund, a mere private, of little immediate interest to the Soviet occupiers at the end of the brief period of hostilities, made his way with his wife and child to the German side of the Polish partition and from there escaped to the United States in March 1940. But Marcinkiewicz was from the Polish nobility and an officer in the army reserve. The Soviets, intent on destroying Polish military capability for the foreseeable future, rounded him up with thousands of other captured officers and moved them east into prisoner-of-war camps. In April 1940, on Stalin's orders, a mass murder of these prisoners was carried out in the Katyn Forest, near Smolensk.

Wirszup continued to teach at the State Technical Institute into 1941. But in June of that year Hitler unleashed his long-planned betrayal of the Soviet Union, sending his army and air forces streaming east across a broad front, from the Baltic to the Black Sea. Lithuania was soon under German control. The Jews of Wilno, including Wirszup, his wife, and his infant son, were herded into a ghetto. He adopted the guise of a house painter, hiding the dangerous fact that he was an educated man.

By 1943, with the German war effort beginning to falter, the home front relied increasingly on forced labor to sustain military production. Despite increasing deprivation, twenty-eight-year-old Wirszup was still viewed by the Nazis as fit enough to perform hard physical work. This may have spared his life, but it separated him from his family. He was sent to a series of concentration camps across the Reich, from Estonia to the border with Switzerland, according to the whims of the Nazi bureaucracy. At the very limit of his endurance, Wirszup was saved by the kind attentions of French political prisoners, captured resistance fighters, themselves facing the likelihood of being literally worked to death.

At the end of April 1945 Izaak Wirszup was liberated by American forces from the camp at Allach-Dachau, near Munich. He made his way back to Wilno, once again under Soviet control, and found what he had feared: his wife, his son, his parents, his brother, and his sister all had perished at the hands of the Nazis. In Wilno he met another survivor, Pera Poswianksi, an old acquaintance from his youth, whose husband had been murdered by the Nazis. Izaak and Pera would marry, and Izaak would adopt Pera's daughter, Marina. The three moved to Paris, invited by Max Heilbronn, one of the Frenchmen who had aided Izaak during his captivity. Heilbronn, a department store magnate, hired Izaak as a director of research and statistics.

In 1949 Antoni Zygmund invited Wirszup to return to academic life and resume work on a doctorate in mathematics. By this time Zygmund was established as a professor at the University of Chicago, a major figure in what mathematicians admiringly refer to as the "Stone Age." Mathematics Department Chair Marshall Harvey Stone (son of US Supreme Court Justice Harlan Fiske Stone) had expertly raised the department out of a period of doldrums by bringing in a cohort of top-flight mathematicians, notably Shiing-Shen Chern, Saunders Mac Lane, André Weil, and Zygmund. Each is considered one of the great mathematicians of the twentieth century.

Wirszup and his family did indeed go to Chicago, overcoming their prejudice; previously they thought of the city only as the home of Al Capone. While working under Zygmund's direction to earn his PhD, Wirszup taught undergraduates in what was then called the College of the University of Chicago. More informally it was called the Hutchins College, since it was the brainchild of Robert Maynard Hutchins, who had precociously taken the reins as president of the university at age thirty in 1929. Hutchins had tried to revive the ideals of liberal education by combating what he saw as overemphasis on vocationalism and premature specialization. For this he needed a cadre of faculty who would devote special attention to introductory undergraduate teaching.

As soon as Wirszup started teaching Chicago undergraduates, he observed that they were not as well prepared for mathematics as he and his contemporaries had been in Poland. He began to ponder the differences between the training he had received and American grade school education. And he began to research how the tradition of mathematics training was continuing in Eastern Europe. By the time he finished his dissertation for Zygmund, an exercise in pure mathematical research titled "On an Extension of the Cesaro Method of Summability to the Logarithmic Scale," he had made a decision. He would not pursue a research career but instead would devote himself to improving mathematics education in the United States. He saw this as the best way in which he could contribute to the field, given his level of talent. It would be one way of giving back, one offering he could make in response to the miracle of his surviving the Holocaust.

Wirszup and Zygmund had had a frank talk about Wirszup's future. Zygmund had recently spent a year in Argentina on a Fulbright fellowship, and there had discovered his second great student, Alberto Calderón. Calderón, brought to Chicago, would be the one to extend the research program initiated by Zygmund and Marcinkiewicz in the 1930s. Zygmund knew that Wirszup was not a Marcinkiewicz or a Calderón. And so when Wirszup asked advice, Zygmund observed that if Wirszup abstained from research, Wirszup's unwritten papers and unproved theorems would be quickly done by someone else. But the contribution he could make to school education by bringing to bear his knowledge of Eastern Europe was far more likely to be unique.

With Wirszup leading the effort, the College Mathematics Staff won a grant from the National Science Foundation to conduct a survey of Eastern European school mathematics literature. This would blossom into a project, supervised by Wirszup, to translate the best of this literature into English. More than 60 volumes would eventually be translated, primarily from the Soviet Union.

Through the translation project Wirszup became aware of the keen interest of Soviet mathematics educators in the psychology of learning. In particular, the Soviets had studied and extended the psychological theories of the Dutch mathematics educators Pierre van Hiele and Dina van Hiele-Geldof, students of noted mathematician Hans Freudenthal. The van Hieles had emphasized that successful students of geometry progress through a series of levels, from simple visualization of shapes to rigorous logical deduction. Failure to fully ingrain the intuitive aspects of geometry leads students to struggle with more abstract aspects. The Soviet interest in these ideas was unknown to the Dutch educators until Wirszup pointed it out. Wirszup alerted American educators to the van Hiele theories as well, and they continue to be influential.

Wirszup was generally supportive of the curricular innovations of the New Math of the 1950s and 1960s, and he was a consultant on the largest such program, the School Mathematics Study Group. Some of Wirszup's translation project was accomplished under the auspices of the SMSG. But enthusiasm for New Math waned in the 1970s. This so-called back-to-basics reaction resulted in diminished ambitions for mathematics in the schools. In Wirszup's view, this situation was compounded in the early 1980s by the proposals by the administration of President Ronald Reagan to reduce federal spending on education. Wirszup was especially alarmed when the National Science Foundation's Directorate for Science and Engineering Education was slated for elimination. In response, he made forceful presentations before congressional committees, mobilized others of like mind, and managed to preserve the NSF directorate.

Shortly after his congressional testimony, Wirszup led the effort for the University of Chicago to obtain a major grant from the Amoco Foundation for a mathematics curriculum project that would utilize the insights he had garnered from studying the Soviet literature. Thus in the early 1980s, while many mathematics educators were still hunkering down to weather the backlash against New Math, Wirszup was among the few to launch an ambitious new program, the University of Chicago School Mathematics Project (UCSMP), in which Wirszup remained active well after his official retirement from the university in 1985.

Izaak Wirszup, November 29, 2000. Photograph by David L. Roberts

Even before the Soviet Union crumbled, beginning in 1989, Wirszup had critics who doubted that Russian education should be taken as the standard of excellence. But Wirszup never lost his conviction that there was much that the United States could learn from the Russians, no matter the moral and economic deficiencies of their political system. The fact that the United States, in the view of many, "won" the Cold War did not dissuade him from his opinion that his adopted country remained too complacent about education.

By the 1960s Hutchins was gone from the University of Chicago, and the College Mathematics Staff was absorbed into the Mathematics Department, although it was still understood that some faculty would specialize in teaching and some in research. Wirszup saw the utility of the distinction, but as he became more firmly established at the university he came to feel that the great research titans, not only in mathematics, were removing themselves too much from the undergraduates. His wife, Pera, who had become an instructor of conversational Russian, was likewise concerned. Both were disappointed, when their daughter Marina was a student at the university, at the lack of contact she had with full professors.

The Wirszups saw an opportunity to amend this situation beginning in 1971, when they became resident masters of Woodward Court, an undergraduate residence hall. They invited the elite scholars of the university to their elegantly furnished apartment for informal discussions with undergraduates, blending the traditions of European high culture, relaxed American hospitality, and their own unique warmth and friendliness. At first attracting only a few dozen, these affairs blossomed into a beloved campus lecture series attended by hundreds. Distinguished speakers came from near and far, including economist Milton Friedman, novelist Saul Bellow, and historian John Hope Franklin. But no matter how large the audience, they were all invited back to the Wirszup apartment for refreshments. As one admirer observed, "Izaak and Pera presided over a *salon* that was very much out of Proust. The Wirszup's 'events' were the talk of the campus, the city, and the world. They were breathtakingly wonderful as only Pera and Izaak could make them."

After stepping down as masters of Woodward Court in 1985, with Izaak's retirement, the Wirszups remained active in the University of Chicago community. Izaak Wirszup died in 2008, aged ninety-three, and Pera Wirszup in 2015, aged one hundred.

17

Carrying Old Virginny Forward

Edgar L. Edwards Jr.

1960

Nothing very significant is accomplished, really, in offering physics or calculus to rural Negro boys who intend to drop out at the ninth-grade level and go to work farming or cutting pulpwood.

James Jackson Kilpatrick

DURING THE SUMMER OF 1960 Edgar Edwards Jr., a mathematics teacher from a rural high school, enrolled in a class at the University of Virginia. Just being able to set foot on the Charlottesville campus was a small victory. Edwards had been repeatedly rebuffed in his previous applications to study there. He had all the necessary credentials, but he was informed that since the classes he wanted were also offered at inconveniently distant Virginia State College, he would not be allowed to attend UVA, a little more than 20 miles from his home. Both of these institutions were supported by the Commonwealth of Virginia, but one was for whites and one was for blacks.

The new dispensation of 1960 was no change of heart by UVA officials regarding racial integration, but simply another in a series of cautious political calculations. An earlier calculation had occurred in 1950 when Gregory Swanson was admitted to the law school under court order. This was reckoned to be cheaper than creating a blacks-only law school from scratch at Virginia State, an alternative that the court would have allowed at the time.

The admittance of Edwards in 1960 represented a slightly different calculation. The summer class to which he was admitted was an introduction to the curriculum innovations then sweeping the country, often referred to as "New Math." This New Math survey was part of a "summer institute," one of many such all across the country, in both science and mathematics, supported by the National Science Foundation (NSF). Grants were provided to the host universities for the instructors, and stipends were provided to the attendees, most of whom were high school teachers. The key that opened the doors of UVA to Edwards was the requirement that the institutes be open to all races. UVA, eager for the federal grant money, did the absolute minimum to comply, by admitting one black teacher to the science institute and one to the mathematics institute. The mathematics teacher was Edwards. The decision to admit Edwards would have a profound effect on his career, opening many new opportunities while changing UVA in a small but significant way. Thus, in this largely unheralded manner, did New Math influence the cause of racial integration in education.

Virginia had long prided itself on being more refined in its racial bigotry than the states of the Deep South. When the US Supreme Court in 1954 issued its momentous *Brown* v. *Board of Education* decision, declaring school segregation unconstitutional, Virginia became a center of dissent from the court ruling, but the state largely avoided violent incidents. Senator Harry Byrd became a vocal leader of the "Massive Resistance" movement to defy the court. When a few local jurisdictions in Virginia showed an interest in school integration, they were promptly slapped down by the state legislature through threats to withhold funding. Prince Edward County, southwest of Richmond, closed its public schools for five years and facilitated private schooling for white students, rather than make the slightest move to integrate.

One of the primary intellectual advocates of Massive Resistance was James J. Kilpatrick, editor of the *Richmond News Leader*, who in 1962 published a book titled *The Southern Case for School Segregation*. Said case, buttressed with learned references and couched in agile prose, boiled down to the bald claim that whites were a superior race. But Kilpatrick enjoyed posing as a moderate. The *Brown* decision had overturned the reasoning of the 1896 *Plessy* v. *Ferguson* decision, which had been used to support a policy of "separate but equal" educational facilities. Kilpatrick conceded that the "equal" part of the *Plessy* doctrine had been widely ignored, plaintively suggesting that much of the later social upheaval might have been avoided if only more attention had been earlier given to equalization. In saying this, he was momentarily ignoring that his book's whole thrust was that blacks were unworthy of equal treatment.

———

Edgar Edwards had lived with the thorough fraudulence of "separate but equal" all his life. He was born in 1922 in Fluvanna County, Virginia, in the tiny hamlet of Kents Store, population 37. There was no electricity, and automobiles were almost unknown. His father, a farmer, had no education past seventh grade, but his mother had attended the Henderson Institute, a secondary school in her home state of North Carolina. The Henderson Institute had been founded in the 1880s by Presbyterian missionaries seeking to educate and evangelize African Americans in the region. Edwards was one of twelve children. The nine who grew to adulthood all graduated from high school, with strong parental encouragement. Edgar, the youngest, was the only one to attend college.

His elementary school owed its existence to wealthy philanthropist Julius Rosenwald. Beginning in 1913, Rosenwald facilitated the building of thousands of schools for blacks across the rural South by challenging local leaders to match the money he put up. Edwards recalled his Rosenwald school as a solid one-room structure, but lacking indoor plumbing and with oil lamps for illumination.

When Edwards graduated from the elementary school, in the early 1930s, there was no high school for blacks in Fluvanna County. There were four high schools for whites, consolidated into one in 1934. A few black families in Fluvanna County paid a man with a rattletrap vehicle to bus their children to attend school in neighboring Louisa County. In this way, with a 3-mile walk to catch the bus, Edwards was able to attend the Louisa Training School. The name itself, as he soon came to understand, was a rebuke to the aspirations of African Americans. The white overlords deemed them incapable of true intellectual attainment. This was at the crux of the argument between Booker T. Washington and W. E. B. Du Bois. For Du Bois,

Washington seemed overly willing to accede to the "training" model of black education, to the neglect of higher cultural goals.

At the level of a specific school, whatever its name, the education offered ultimately depended on the skills and philosophies of the particular teachers. At the Louisa Training School the small staff tried as best they could to offer an academic program for those interested. There was a principal, who also taught English, a math and science teacher, a history teacher, and a French teacher. Looking back, Edwards thought that the quality of the teachers was remarkably high, which he attributed to the dearth of other opportunities for educated African Americans. When those opportunities increased, the schools began to face challenges attracting good teachers, as he would learn firsthand when he became an educational administrator.

There were other forces at work that restricted the scope of Edwards's early education. Schools for both white and black in rural Virginia in the 1930s opened late in the fall and closed early in the spring, to allow the children to do farmwork. A seven-month school calendar was considered generous. The Louisa Training School, when Edwards started there, was offering three years of instruction in three rooms for about 80 students. Here again there was no indoor plumbing. There was no cafeteria and no library. In mathematics, Edwards had a year of eighth-grade general mathematics, a year of algebra, and a truncated year of geometry. That last year the school board of Louisa County consolidated the black junior high with the training school. With the resulting overcrowding, the geometry became a self-study class, with students forced to sit outside in a bus. Edwards did not feel that he was displaying any special facility in mathematics, but he enjoyed learning it, as he enjoyed all his school subjects.

After graduating from the Louisa Training School, in 1939, he mostly lived at home in Kents Store doing farmwork with his father. This was a life that never much appealed to him. Fortunately, a young man happened to be lodging with the family that year, a teacher in the Fluvanna County elementary school that Edwards had once attended. Such teacher hosting was common in rural areas, especially for black teachers. This teacher was a graduate of Virginia Union University, in Richmond, and when he saw that Edwards showed signs of being academically inclined, he helped him to gain admittance to Virginia Union by writing a letter of recommendation.

Somehow Edwards cobbled together the funds needed to enroll. He earned a little money working in Washington, DC, over the summer, living with one of his sisters. Virginia Union, as reflected in its name, had been created from the merger of two schools, theological seminaries founded after the Civil War by Baptist missionaries intent on nurturing the spiritual and educational needs of those newly freed from bondage. Part of the land on which it stood in Richmond had once been the site of a holding pen for slaves waiting to be sold. By the time Edwards arrived in 1940 it was a coeducational institution of about 375 students, with a continuing emphasis on preparation for the ministry. Edwards himself had some thoughts for a time of becoming a preacher. But his initial time at Virginia Union was dominated by the realization that his high school classes had not fully prepared him for college. He needed to take two noncredit classes in mathematics and one in English before entering into the regular curriculum. This experience remains a common one for many students to the present day, especially those from reduced economic circumstances.

Edwards stayed at Virginia Union only one year. The shock of France falling to the invasion of Hitler's Germany in 1940 led the United States to institute its first ever system of peacetime conscription for young men, which was administered by local draft boards. Edwards thus found himself at the mercy of old white men who had served in the Spanish-American War and in World War I. The board was in the powerful position of being able to order young black men to work for minimal wages on construction projects, dubiously justified as defense work, in return for which the board would generously defer them from the draft. The motivation for Edwards and others to agree to this arrangement became even stronger when the United States became an active participant in the war, after the Japanese attack on Pearl Harbor in December 1941. Thousands of draftees died in foreign lands. In later years Edwards was sensitive about his draft deferrals.

The advent of the war also meant that local draft boards came under more outside scrutiny. Soon Edwards's board was no longer able to confine him to Virginia but was obliged to assign him to high-priority projects elsewhere. He was sent to work in Upstate New York, digging tunnels for one of the aqueducts supplying water to New York City. Here the pay was better, he was able to read a daily newspaper, and he was for the first time out from under the constant oppression of the Jim Crow laws in the South. He visited the great metropolis and was awed by the tall buildings, the bright lights, and the subway. He visited museums and saw professional sports teams play.

He continued doing tunnel work after the war, hard physical labor as a sandhog in New York and Pennsylvania, but the pay was so good that he was reluctant to give it up. Whenever he visited his parents in Virginia, they asked if he was planning to go back to school, and finally, in 1947, he returned to Virginia Union. His earlier thoughts of becoming a minister dissipated, as he found that he especially enjoyed mathematics. He discovered he was good at tutoring fellow students who were struggling, and he strengthened his own knowledge in the process. In 1950 he graduated with a bachelor of science degree in mathematics.

Edwards returned to Fluvanna County, and in the fall of 1950 he became a teacher at Samuel Abrams High School. The county, with pressure and funds from its black community, had finally built a high school for black students, just as Edwards was graduating from the school in the neighboring county. It was a four-room school for about 150 students. It had no indoor plumbing and no cafeteria, but it did have a library.

As was to be expected in a small rural school at this time, Edwards taught multiple subjects. In addition to mathematics, he was qualified, on the basis of his college courses, to teach science and Latin. He taught science for several years. There were never enough students to have a Latin class, but his presence on the faculty as a qualified Latin teacher helped the school to maintain its accreditation with the state. As with all teachers at the school, he taught health and physical education. He also coached basketball, football, and baseball. As the enrollment grew, the school added more classrooms and more teachers, and Edwards became more of a mathematics specialist, while continuing to teach health and physical education. He taught general mathematics, algebra, geometry, and trigonometry.

Edwards was teaching at the Abrams High School in 1954 when the Supreme Court handed down the *Brown v. Board of Education* decision. It changed his life hardly at all. Fluvanna County played a waiting game, proceeding with minimal change until the mid-1960s.

Another well-publicized event of the 1950s had a more immediate effect on Edwards's career: the 1957 launch by the Soviet Union of Sputnik, the first artificial satellite to orbit the earth. Edwards, like many Americans, was amazed by this technological feat, and he noted the ensuing complaint that the American failure to be first to achieve this advance indicated deficiencies in science and mathematics education. He heard about new topics being introduced into the curriculum. It was at this time that he started to apply for graduate study at the University of Virginia. On being rebuffed there, he gained admittance to an NSF institute in the summer of 1959 at George Washington University in Washington, DC, where he took both abstract algebra and calculus. The former represented New Math, while the latter represented an advanced level of the old math, increasingly being taught in twelfth grade in Virginia high schools. It had been over a decade since Edwards had had calculus, and he felt in need of a refresher.

Taking the classes at George Washington, Edwards could stay with his sister in Washington, but it was still not ideal, as he now had a wife and family back in Fluvanna County. So Edwards continued to apply to UVA, and finally in 1960 he was admitted. The graduate school then agreed that if his grades in the summer were no lower than a B-plus average, then the school would support him through an academic year and the following summer. As a result, Edwards quit his teaching job and became a full-time graduate student, allowing him to earn a master's degree in education in 1961. Slowly but surely, UVA was beginning to truly integrate.

At UVA, Edwards was especially impressed with a class in the foundations of geometry, taught by professor Billy Joe Ball. Ball taught with an unusual technique, the significance of which Edwards would not fully understand for many years. Ball did not lecture, and there was no textbook. He began by stating some undefined terms and some axioms, and then proved a couple simple theorems based on these axioms. Then he gave the students a list of further theorems, and from then on it was up to the students to do the proving. A student would present a proof at the board, critiqued by the other students or by Ball, if the students were not sufficiently critical. All proofs had to be based on the axioms or on previously established theorems. There were some mathematics doctoral students in the class, and Edwards noticed that they would sometimes walk out when a student at the board was starting to prove a theorem that the doctoral students had not yet proven on their own.

Edwards found this a stimulating course, but he never had the opportunity to investigate its provenance or to employ its methods himself, because the master's degree he obtained from UVA redirected his career in mathematics education out of classroom teaching. He had expected to return to the high school in Fluvanna County, to initiate it into the new curricular topics and possibly offer calculus. But he found, to his great shock, that the principal refused to rehire him. Later, Edwards came to realize that from the principal's point of view, Edwards's master's degree looked like a ploy to take the principal's job. Edwards was briefly unemployed, but by coincidence the state of Virginia had created a new position supervising mathematics in black high schools, with a particular view to incorporating some of the New Math. Edwards, with his recent degree, was ideally qualified. Moreover, the state supervisor for mathematics, Isabelle Rucker, had attended a UVA summer institute shortly before Edwards. Through her consequent acquaintance with UVA professors she was able to get reports on Edwards's performance at UVA, thus contributing to his being hired in Richmond as assistant supervisor of mathematics for Negro high schools. In retrospect it is clear that this was a belated attempt to provide some semblance of equality in the still largely segregated system, with a view to delaying full integration as much as possible. At the time, it was too good an opportunity for Edwards to pass up.

Edwards was not the only schoolteacher to use an NSF summer institute, designed to enrich his schoolteaching, to instead vault out of schoolteaching. Such an unanticipated consequence is a continuing issue with all so-called in-service training in the United States and speaks to the hierarchy of prestige, and compensation, within the educational system. Teachers are markedly inferior to administrators, and schoolteachers are markedly inferior to college teachers. Giving a schoolteacher an added credential may simply allow that teacher to escape the school.

At first, Edwards's job with the state was largely segregated. He worked with the black schools, and his boss, Rucker, worked with the white schools. Every year there were two state conferences for mathematics teachers, one for blacks and one for whites, usually on the same weekend but at different locations. Edwards and Rucker would meet back at the office to discuss how the conferences went.

Much attention was focused on accreditation of black schools by the state and the Southern Association of Colleges and Schools. Here again there was a late push to give a veneer of reality to "separate but equal." Accreditation needed to be removed as an issue for litigious black Virginians. Edwards was advised to drop all other matters if a black school needed propping up to pass accreditation. He held his tongue and soldiered on, feeling that at least he was helping to create better schools, waiting for integration.

Things began to change in the mid-1960s, as massive resistance finally began to wane and integration became official state policy. By no means did this create truly integrated schools, but it did mean that no school was designated as white or black. Edwards was now simply the assistant supervisor of mathematics for high schools, and he found himself visiting all sorts of schools, some predominately white. More than once a suspicious white principal would ask, "You're not going to start anything here, are you?" He was increasingly called in on issues not related to mathematics, as a sort of general consultant on racial etiquette. How should the school lunch menu be set up to accommodate black students attending a formerly all-white school? How should pictures of whites and blacks be arranged in the hallways?

Edwards was happy to resolve such issues, while wishing he could devote himself more fully to mathematics. In 1973 he succeeded Rucker as the overall supervisor of mathematics for Virginia, the first African

American to hold this position. Increasingly, he had to deal with the strong feelings evoked by education reform. New Math had lost its impetus, with many calls across the nation for a return to the "basics." It was a ticklish issue to decide what those basics were. Edwards questioned, for example, whether division of fractions needed as much emphasis as traditionally given. He would attend workshops for teachers and ask if they had ever had a practical need to divide fractions, other than solving a problem out of a textbook. Inevitably, someone would claim that they had needed to do this in preparing half of a recipe. Edwards would patiently point out, to the consternation of the teacher, that this was multiplying by one half, not dividing by one half. "I had to quit asking the question, because they would get angry with me." He also learned how negatively many schoolteachers react to an article or talk that begins with "Research shows that . . ." Nor are teachers always enthused to hear about how education is done in other countries.

In the 1970s Edwards became prominent in the National Council of Teachers of Mathematics, which he had originally joined in the late 1950s. In 1994 he would be awarded the NCTM Lifetime Achievement Award. He served on committees and eventually was elected to the board of directors. In 1986 he was a founding member of an organization associated with NCTM, the Benjamin Banneker Association. Named for the eighteenth-century surveyor and almanac maker, one of the first African American mathematical practitioners, this group advocates for "leveling the playing field for mathematics learning of the highest quality for African-American students." In 1990 Edwards was a major proponent of NCTM's "Algebra for Everyone" initiative, which sought to raise the level of mathematical learning for all American students. This converged with the Algebra Project, led by Robert P. Moses, veteran of the 1960s voting registration struggle in Mississippi, who framed high-quality mathematics education as a civil rights issue.

In 2002 Edwards, then living in retirement in Richmond, was interviewed for an NCTM oral history project, during which time he finally learned about the lineage of the UVA geometry course he had taken more than 40 years before. The instructor, Billy Joe Ball, had been a doctoral stu-

Edgar L. Edwards Jr., October 17, 2002. Photograph by David L. Roberts

dent at the University of Texas under the famed mathematician R. L. Moore, and Ball's distinctive method of teaching had been pioneered by Moore. In the late 1990s intellectual descendants of Moore had begun a concerted effort to bring this "Moore Method" of mathematics teaching to a wider audience, connecting it with other ongoing pedagogical offerings under the general title of inquiry-based learning. Edwards was put in contact with the Moore Method devotees. As a consequence, in March 2003, two years before his death, Edwards was invited to the Sixth Annual Legacy of R. L. Moore Conference in Austin, Texas, where he recounted his memories from the early 1960s. He also learned of the irony that R. L. Moore was well known for being an unvarnished racist who had refused to tolerate anyone with dark skin in his classes. But Moore's students, among them Ball, although admiring of their mentor's mathematical and pedagogical accomplishments, had emphatically broken with his racism. Two of the first three black Americans to earn a PhD in mathematics were doctoral grandsons of Moore.

18

Americano

Joaquin Basilio Diaz

1974

It occurs to me, and this is a comment on the manner many mathematical papers are written nowadays without being read by anyone except the authors themselves, that perhaps, instead of "everything I will say is contained in these three papers," I should have said "everything I will say is very carefully hidden in these three papers."

J. B. Diaz

JOAQUIN BASILIO DIAZ, known as "Joe" to his friends, published his one hundredth mathematical research paper in May 1974. It seemed to Diaz that this milestone was worthy of being memorialized, so he set about compiling a complete annotated list of his publications. To this list he appended all relevant entries from *Mathematical Reviews* and *Zentralblatt für Mathematik*, services established by the international mathematical community to review significant research papers. He also attached an inventory of his 23 PhD students, including for each one the dissertation title and subsequent career trajectory. It took him three years to assemble all this information, by which time, because his mathematical productivity had continued undiminished, he had published thirteen more papers. Mathematical colleagues around the world received copies of the resulting document, which Diaz called his "apologia pro vita mea." The following year, in June 1978, having published one more paper and having guided one additional graduate student to a doctorate, Diaz died suddenly, age fifty-eight.

Although he was not the founder of a "school" of mathematical research, he influenced a substantial contingent of mathematicians. The list of publications in his "apology" shows that he was generous with intellec-

Joaquin Basilio Diaz, 1974. Photograph by Richard C. Roberts, in possession of David L. Roberts

tual credit, typically writing in collaboration with others, especially with his students. Several of his papers include an acknowledgment that research support was provided by one of the military services, underlining the fact that Diaz was an applied mathematician whose career was shaped by the Cold War between the United States and the Soviet Union. He was one of the first mathematicians of Hispanic heritage to get a PhD in the United States, although such a distinction would have meant little or nothing to him. He would have preferred to be remembered as a good citizen of mathematics.

Diaz was born in in 1920 in Arecibo, on the north coast of Puerto Rico, about 50 miles west of the capital, San Juan. His parents did not have deep roots in this Caribbean island. His father had in 1913 emigrated from Málaga, Spain, born there in the 1880s just a few years after that city's most famous son, Pablo Picasso. Diaz's mother, although born in Puerto Rico, had a father who had also come over from Spain. For over 400 years, from the second voyage of Christopher Columbus until the end of the nineteenth century, Puerto Rico had been a Spanish colony. But in 1899, after the flourishing United States crushed feeble Spain in the brief conflict known as the Spanish-American War, Puerto Rico had become a territory of the United States. So it has remained to the present day. Occasional tweaks to the administrative rules have not removed the essential anomaly of the relationship, which continues to perplex politicians on both the island and the mainland.

From the first days of American control in Puerto Rico, there were protests against territorial status. During World War I, the US military became alarmed at the prospect of German incursions into the Caribbean, and consequently more sensitive to political unrest in the region. In 1917, with American entry into the war looming, Congress and President Woodrow Wilson agreed on a plan to counter independence agitation on the island by granting US citizenship to Puerto Ricans—or, as Puerto Rican nationalists viewed it, imposing US citizenship. For the military it created a new pool of potential draftees for service in the war. For Joaquin Diaz, it meant that he was born a US citizen.

His father was a dry-goods merchant and his mother a seamstress. They both spoke and wrote only Spanish all their lives, but young Diaz, with others of his generation in Puerto Rico, began to learn English as soon as he started school. Already in the census of 1930, when he was nine years old, he was reported as being literate in English. He would, however, always speak it with a strong accent. After living for many years on the mainland, he discovered that Puerto Ricans on the island identified him as an "Americano" when they heard him speak Spanish on the telephone.

In high school Diaz was a superb student, shining especially in mathematics. Seeking to cultivate his scholarly excellence, his family sent him at age sixteen to attend college on the mainland. Following the advice and example of three students from Arecibo who had preceded him, he enrolled at Washington and Jefferson College, a small, all-male, liberal arts school 30 miles from Pittsburgh, Pennsylvania. Here he stayed two years. The ship voyages he took between San Juan and New York City at the beginning and end of each academic year demonstrated to him the limits of his status as a US citizen. He was listed among the "alien" passengers, and his nationality was recorded as "Spanish." Border-crossing indignities would vex him for the rest of his life.

After learning of the innovative mathematics instruction at the University of Texas (see chap. 17), Diaz transferred to the Austin campus in 1938. He graduated with a degree in pure mathematics and was elected to Phi Beta Kappa, the academic honor society. He then continued at Texas for graduate study. He was briefly fascinated by point-set topology, the mathematical subfield specialized in by the leader of the Texas pure mathematicians, R. L. Moore, a fixture on campus since 1920. Topology was an extension of geometry that emerged vigorously in the first half of the twentieth century, with many Americans at the forefront. Originally referred to as the geometry of position, or as analysis situs, topology is concerned with those properties of geometric figures (or sets of points) that do not depend on the familiar concepts of length and angle measurement. Moore was promoting an especially austere version of topology, sharply circumscribing its connections with other parts of mathematics. But students who flourished under his powerful personality and his distinctive teaching regime, with independent problem solving rigorously enforced, became impressively resourceful mathematicians. He turned out dozens of them. Moore was also a strong supporter of the University of Texas policy of excluding black students. As one former student observed, Moore at times looked askance at persons without Anglo-Saxon names as not being "really American." In the long run, Diaz could not thrive in this atmosphere, although fellow students remembered him fondly.

Before he left the University of Texas, Diaz established a close relationship with one professor, Hyman Ettlinger, who specialized in an area of mathematics more traditional than topology: differential equations. Seeking to expand his knowledge of this field, Diaz traveled to Providence, Rhode Island, in the summer of 1942, attracted by Brown University's

Program for Advanced Instruction and Research in Mechanics. This program had begun the previous summer, as Brown began a concerted effort to augment its offerings in applied mathematics, an effort that seemed more urgent with the entry of the United States into World War II in December 1941. Courses were offered on the theory of flight, on fluid mechanics, and on electromagnetic waves, all topics to which differential equations could be profitably applied. These summer programs would lead to the formation of Brown's Division of Applied Mathematics, officially established in 1946. Taking advantage of a pool of brilliant mathematicians expelled by Hitler, and augmented by eager young Americans, Brown rapidly became a premier center for applied mathematics in the United States. This was counter to what had been the general trend of mathematical research in the country, dominated by theorists and aesthetes since the beginning of the twentieth century. Nathaniel Bowditch's early nineteenth-century characterization of Americans as preferring practical things to theory (see chap. 1) had long been untrue in mathematics.

Diaz found his summer of applied mathematics tremendously invigorating. It led to his enrolling at Brown as a graduate student, under the guidance of a mathematician strikingly different from R. L. Moore: Lipman Bers. Just six years older than Diaz, Bers was a Latvian Jew who, after earning a PhD in Prague and spending a brief interlude in Paris, had barely escaped Europe ahead of the Nazis. He had an expansive view of mathematics, moving easily between pure and applied topics. In Europe, Bers had been outspoken on social justice, and he continued his activism in the United States. He welcomed cultural diversity. The international language of science, Bers quipped, was "heavily accented English."

Diaz, with a 1945 dissertation on partial differential equations, became the first student to receive a PhD under Bers's direction, a distinguished group that would eventually number more than 50. Even in his youth Diaz had been interested in the history of mathematics, and later in life he became fascinated by mathematical genealogy, the sequences created by the doctoral advisor-advisee relationship. He was proud to note that through Bers he could trace his mathematical forebears back to one of the giants of nineteenth-century mathematics, Germany's Karl Weierstrass (1815–97).

In recent years many mathematicians have become similarly intrigued by such connections, an interest now embodied in the popular website of the Mathematics Genealogy Project.

Bers also further energized Diaz's already considerable interest in languages. Bers had impressed upon Diaz that a wealth of mathematical research was written in Russian, in which Bers was fluent. Consequently, Diaz added that language to his repertoire, which already included French, German, and Italian in addition to Spanish and English. He became so expert in Russian that he translated many papers from Russian to English for the American Mathematical Society.

From the time he arrived at Brown, with World War II at its height, Diaz was vulnerable to conscription into the US Armed Forces. Expecting to he drafted, he and a graduate school colleague temporarily deafened themselves while taking instruction on shooting handguns from a marine sergeant. But Diaz remained a civilian throughout the war. It helped that from the middle of 1943 to the middle of 1945 Bers hired him as a research assistant, working on a contract for the National Advisory Council for Aeronautics, the predecessor agency to the National Aeronautics and Space Administration. The NACA work concerned compressible flow. At the theoretical level it represented another excuse to play with differential equations, but Diaz also developed skills in making the judicious computational approximations that characterize the effective applied mathematician. He would use these skills in consulting work for much of his later career.

The military had used a substantial amount of mathematical talent during World War I, notably to study ballistics problems at the Aberdeen Proving Grounds in Maryland. The World War II mobilization was far more extensive, with the Manhattan Project, the crash program to build an atomic bomb, being only one among many military employers of mathematicians. As a consequence of the Cold War, there was no pronounced drawdown after the 1945 armistice, as there had been in 1918. Many mathematicians were engaged to work for the military in the 1950s, either directly by the government or as contractors. Only a small minority of politically radical scientists and mathematicians had any qualms about such connections, until the reaction against the Vietnam War in the 1960s.

For a time, budgets were sufficiently lavish that the military was agreeable to supporting a considerable amount of basic research. There was always a chance the abstractions might prove useful, and in the meantime a stable of talent could be maintained during the long, slow burn of the Cold War, talent that could be mobilized in a more instrumental manner should a more pressing crisis erupt. Both the air force and the navy supported portions of Diaz's theoretical research on differential equations during the 1950s and 1960s. He spoke at unclassified conferences sponsored by the army and the air force.

At the same time, Diaz did genuine applied mathematics for the government, much of it classified. He had long-term consulting relationships, first with the Atomic Energy Commission (AEC) and then with the Naval Ordnance Laboratory (NOL). The AEC was a civilian agency with close ties to the Department of Defense, established after World War II to manage the facilities created for the Manhattan Project. NOL was concerned with developing all sorts of ship-launched weaponry: mines, missiles, torpedoes. AEC and NOL were both located not far from the University of Maryland, where Diaz worked for 16 years.

Diaz had come to Maryland in 1950, after briefer teaching positions, first at the Carnegie Institute of Technology and then back at Brown. The Carnegie stint was only one year, but it made a strong imprint on the rest of his career, for it was here that he first met Alexander Weinstein, a major researcher in differential equations and their applications. Weinstein, like Lipman Bers, was a European émigré. Born in Russia in 1897, Weinstein had studied in France, Germany, and Switzerland, impressing some of the greatest mathematicians of the early twentieth century, notably Hermann Weyl and Jacques Hadamard. Diaz and Weinstein wrote two joint papers while together at Carnegie Tech, beginning an association that would last, with many ups and downs, to the end of Diaz's life. Their most sustained connection occurred at the University of Maryland, where from 1950 to 1966 they both held appointments at the Institute for Fluid Dynamics and Applied Mathematics. IFDAM was funded largely by the navy, with close ties to the nearby Naval Ordnance Laboratory. Institute researchers frequently taught courses in the university's mathematics department. Diaz taught graduate courses, rarely encountering undergraduates. Thus it was in many ways a desirable place to work for an active researcher such as Diaz. But Weinstein's demanding and difficult personality wore on Diaz, contributing to his decision to leave in 1966. Nevertheless, Diaz in 1977 acceded to one more major favor for Weinstein, editing a volume of the latter's selected papers.

A major favor that Weinstein did for Diaz was to introduce him to collaborative research, which became Diaz's standard procedure. Well over 80% of Diaz's research papers were written with coauthors. For his students, these papers helped launch their academic careers. This was just one of the ways in which Diaz demonstrated his devotion to keeping the gears of the mathematics discipline functioning smoothly. His translation work has already been noted. He did not shrink from the often thankless tasks of being an editor and peer reviewer for mathematical journals, and he was a faithful attendee and speaker at meetings and conferences, nationally and internationally. The increasing fragmentation of mathematical knowledge bothered him, with leading specialists content to speak only to a tiny clique of those in the know. At meetings he made a concerted effort to make his talks inviting, with puns and witty remarks. He delighted in having his listeners "rolling in the aisles." His reputation for entertaining presentations was such that at sessions with multiple speakers the audience would swell markedly just before his scheduled talk, and then disperse before the next speaker on the program.

By the mid-1960s, Joe Diaz was a distinguished but not towering figure in American mathematics. It was a boom time for American colleges and universities generally. New campuses were springing up, existing campuses were expanding, and competition was growing for top-flight professors. Salaries increased; teaching loads decreased. Diaz, after staying with one institution for 16 years, moved twice in 11 months, with his wife and two children.

The first move, in 1966, was to the University of California, Riverside, an ambitiously growing campus within that state's sprawling system. The warm weather appealed to Diaz, who had been missing the climate of his native Puerto Rico. Initially he was promised a position where he would again be primarily a researcher, not an administrator. He took several of his doctoral students with him from Maryland. But when he arrived, the mathematics department insisted on making him chairman. It was not a happy situation. He soon began to look for employment elsewhere, with climate conditions becoming entirely secondary. Toward the end of the academic year he surprised his two children by making faces in the mirror and cryptically inquiring, "Do I look like Albert Einstein?" The mystery was solved a few weeks later when he announced that he had been appointed to the Albert Einstein Chair of Science at Rensselaer Polytechnic Institute (RPI). The family would be moving back east, to New York State.

RPI was located in Troy, on the Hudson River, just north of Albany. Founded in 1824, it was a pioneer in American engineering instruction, preceded only by the US Military Academy at West Point (see chap. 2). In accordance with the general convergence of twentieth-century higher educational institutions, by the 1960s RPI looked much like other universities of its size, while maintaining an engineering emphasis. In 1967 the chairman of the RPI mathematics department was George Handelman, who had been a fellow graduate student in applied mathematics with Diaz at Brown in the 1940s. By the 1960s Handelman had become a politically astute academic administrator, just the man to take advantage of a funding opportunity offered by the state of New York.

In 1964 the New York State Legislature had caught the attention of the academic world by seeking to attract elite professors to universities in the state through the creation of ten "super-chairs," five in the sciences and five in the humanities. These were designated, respectively, Albert Einstein chairs and Albert Schweitzer chairs, a symmetry evidently pleasing to the legislators, although neither of the two eminent gentlemen being memorialized had much connection to New York. But the most eye-catching feature of the legislation was that each chair was funded at the then-staggering sum of $100,000 per year. This figure would be mentioned in all media reports about the chairs, not always taking care to clarify that less than half this amount was going to the individual chair holder. The money was also used to cover such expenses as office supplies, secretarial assistance, laboratory equipment, conference organizing, and fellowships for graduate students.

Some highly distinguished individuals were installed in these chairs, spread around the state at both public and private universities, as stipulated by the legislature. Historian Arthur Schlesinger Jr., not far removed from his well-publicized service in the White House under President John F. Kennedy, was appointed to a Schweitzer chair at the City University of New York. New York University gave its Schweitzer chair first to diplomat and writer Conor Cruise O'Brien, and then to novelist Ralph Ellison. C. N. Yang, a Nobel Prize–winning physicist, was given an Einstein chair at the State University at Stonybrook. And Joaquin Diaz was named to the Einstein chair at RPI.

Conditions for Diaz at RPI were excellent, at first. He pursued his research, mentored a flock of graduate students, and continued his service activities for the wider mathematical community. But the boom times for academia came to an end in the early 1970s. Campus demonstrations related to racial issues and the Vietnam War helped make politicians less friendly to universities. In 1971 the New York legislature began to cut the funding for the super-chairs. Diaz began to joke that his chair was in danger of collapsing underneath him. Feeling political heat, RPI required Diaz to teach undergraduates in addition to graduate students.

The sudden death of Diaz in 1978, from a heart attack, was a shock to all who knew him, although afterward some recalled that he had expressed an aversion to living to an advanced age. At the time of his death he was in the midst of planning another research collaboration, with Deborah Tepper Haimo, at that time a professor at the St. Louis campus of the University of Missouri. Their joint work was to be on the heat equation, one of the most famous partial differential equations in mathematical physics. Although it had been extensively studied since the 1820s, Diaz was confident that something new and clarifying could be done. Such was his typical attitude to his field.

RPI memorialized Diaz with a prize in his name, to be awarded "to a graduate student who shows ability and enthusiasm for research in mathematics." RPI also gratefully accepted nearly 1,000 books that Diaz had owned, donated to the school by his widow. In 1986 the Diaz Prize winner for that year, who had been born in the Soviet Union, wrote to Mrs. Diaz expressing how much he had valued the Diaz book collection. He noted the many Russian books once owned by Diaz, not only in mathematics but also in nineteenth-century poetry.

The Diaz Prize has continued to be awarded at RPI, but as those who knew him personally have passed away, his name has appeared less often in mathematical discourse. There is no Diaz theorem or Diaz method or Diaz program. He did not astonish with path-breaking results or ambitious agendas for research. Rather, he sought to reexamine basic theorems and their consequences, to simplify proofs, to explore a familiar problem or phenomenon from a new direction and thereby come to improved understanding. He wrote more than a dozen papers on aspects and extensions of the "mean-value theorem," a fundamental result encountered in any course in the differential calculus. If a smooth curve (one with no sharp corners) is drawn between two points, then somewhere between those two points the tangent to the curve will be parallel to the straight line drawn between the two points. As he remarked in a talk he gave in 1974, his ultimate concern was with "the gradual process of unification and simplification which, though usually little publicized, is so necessary for the continuing progress of mathematics." This attitude is not universal among mathematicians. Of Paul Cohen, a twentieth-century mathematical superstar, it was observed that "He looks down on mathematicians who do mathematics for the sake of making incremental improvements in the field."

In the 1990s, about 15 years after his death, there was a flurry of interest in raising Diaz's visibility, but it led to an awkward denouement. By that time, the relative lack of participation in mathematics by African Americans and Hispanic Americans had become an increasingly debated political issue. In 1993 the Mathematical Association of America (MAA) announced a plan to honor an African American and a Hispanic American for their contributions to mathematics, by naming rooms in their honor at the MAA headquarters building in Washington, DC. In addition to publicizing role models to attract future mathematical practitioners, this was a fundraising device for the MAA, a professional organization focused on the undergraduate teaching of mathematics. Potential donors were solicited to give money to the MAA in memory of the individual to be honored with the room. The African American chosen was Benjamin Banneker (1731–1806), a self-taught astronomer and surveyor who worked on laying out the boundaries of the District of Columbia. The Hispanic American chosen was Joaquin Basilio Diaz.

The campaign for the Banneker Room was uncontroversial. The target goal of $15,000 was met and the room duly dedicated, with a plaque by the door. But the Diaz Room never came to reality, highlighting the delicate nature of even the most well-intentioned effort by a privileged group, from its lofty height, to honor those seen as less privileged. Diaz's widow, Eleanore, was sent a copy of the MAA newsletter in which the room-naming campaign was announced. The announcement referred to Diaz as a "minority mathematician" and a "Puerto Rican mathematician." Eleanore Diaz took strong exception to both these designations, explaining that her husband had considered himself simply an American mathematician, "completely integrated into the Mainland of students, professors, and mathematicians." The MAA felt that they could not conscientiously promote the Diaz Room without touting him as a "minority," and so the campaign was quietly dropped.

All evidence suggests that Mrs. Diaz was correct, and that Diaz felt no special affiliation with other mathematicians of Hispanic heritage. Other American mathematicians, both older and younger than Diaz, have been insistent on their Hispanic identity. This was not his way. But as a partial listing of his doctoral students testifies, he launched the careers of mathematicians from an impressively wide range of backgrounds, including Beatriz Margolis (from South America), Mohammed T. Boudjelkha (from northern Africa), Vaclav F. Pfeffer (from Eastern Europe), Kadir Aziz (from Afghanistan), Ram Bachan Ram (from India), and Wallace Parr (of Native American heritage).

19

Math Warrior

Frank B. Allen

1984

> But even these well-defined and successfully taught programs could not withstand the destructive forces generated by the times and mobilized by a handful of critics whose views were expressed vociferously and hyped by the media.
>
> *Frank B. Allen*

I N THE NOVEMBER 1984 issue of the *Mathematics Teacher*, Frank B. Allen published "The New Math—An Opportunity Lost," lamenting the decline of mathematics education in the United States since what he considered the golden age of the early 1960s. Allen had been a major participant in the reform efforts of that era and had served as president of the National Council of Teachers of Mathematics (NCTM) from 1962 to 1964. He attributed the decline of the promising curricular programs to unfair derisive criticism, led by applied mathematicians, combined with a public declaration of loss of faith by an especially prominent reformer. The situation became even worse, according to Allen, when an anti-intellectual counterculture rose up on college campuses in the late 1960s, resulting in a disparagement of science and technology and a turn away from rationality in general.

Allen's piece appeared in a regularly appearing section of the *Mathematics Teacher* called "Soundoff," which explicitly encouraged provocative views. But Allen's hope for sparking vigorous debate was disappointed. He had honed his polemical skills in the preceding decades and was looking forward to eviscerating fuzzy-thinking opponents. But mathematics education in the 1980s was enjoying a lull in public political controversy.

Hardly anyone bothered to reply to Allen. It was not until the 1990s, with the advent of what came to be called the "Math Wars," that a new eruption occurred. Professors, teachers, parents, and government officials were all drawn in, and Frank Allen, although by then in his eighties and completely retired, finally had an opportunity to again broadcast his firmly held views.

———

Allen was born in 1909, and his experience with mathematics education went deep, almost all of it in the state of Illinois. He started teaching before his twentieth birthday. He had attended a series of schools in the rural part of the state, as his father, a Presbyterian minister, was transferred from church to church. After his father died in the flu epidemic that swept the country and the globe after World War I, Allen and his mother settled in Carbondale, where he graduated from the University High School connected with the Southern Illinois State Teachers College (later Southern Illinois University). He then continued on at the college; tuition was low for those, like Allen, who made a commitment to teach after graduation. Though initially interested in both mathematics and physics, he gravitated toward the former, finding that he did not enjoy the laboratory work in physics. He also realized that most high schools had at most one physics teacher, while employing several mathematics teachers.

That first teaching job, beginning in 1929, was at the high school in Sparta, a small town in southwest Illinois. Here he had an opportunity to teach a variety of classes: plane and solid geometry, algebra, trigonometry, and financial mathematics. The firm hand of the principal and the cooperative attitude of the parents in supporting disciplinary measures produced a tightly run school that Allen would look back on as an ideal to be emulated. Allen and the town managed to weather the depression of the early 1930s. During the summers Allen earned a master's degree in mathematics at the University of Iowa. Later in the 1930s he took additional graduate work in mathematics at the University of Illinois, without any further degree.

After seven years at Sparta he moved on to other high schools. He advanced north, first to Urbana, then to Calumet City, and just as World War II broke out, to Lyons Township High School and Junior College in Lagrange, a western suburb of Chicago.

Allen looked back on many of his war experiences with wry amusement, always aware of how fortunate he had been to have spent the entire conflict stateside. Drafted into the army in 1942 as a private, his level of education sent him to officer training school in Fargo, North Dakota ("only a barbed wire fence between there and the North Pole; it was cold!"). He was then transferred to Oklahoma as a newly minted second lieutenant ("the lowest form of human life in the army"), where, as a total novice with weaponry, he was assigned as a small arms instructor for a cadre of combat-hardened veterans who were temporarily back from fighting in Europe ("a dire situation"). Eventually the army rectified this state of affairs:

ARMY: We have a position here for an insurance officer. Are you
    interested in insurance?

ALLEN: No, I'm not.

ARMY: Do you know anything about Army insurance?

ALLEN: No, I don't know anything about it.

ARMY: You're the insurance officer.

Later he was assigned to Summary Court, adjudicating cases where enlisted men were accused of petty offenses. He spent the rest of the war involved in the military justice system. Some observers claimed to glimpse the need for reforming mathematics education in the behavior of the troops during the war, but Allen was unable to see much relevance. He did, however, feel that his mathematical training in proof and logical reasoning had enabled him to navigate the legal cases that he was asked to deal with in the army. This buttressed his strong conviction that studying mathematics could benefit many students in later life, even if they never used the subject directly.

On being released from service Allen returned to Lyons Township High School and Junior College. Aside from the adjoined junior college (a feature specific to some Chicago-area high schools from the early twentieth century to the 1960s), Lyons was an archetypal educational institution of the postwar era in affluent white suburbia: a large comprehensive high school. By the mid-1950s there were more than 2,400 high school students at Allen's school, with most proceeding to college after graduation. But despite such a relatively homogeneous population the issue of individual differences emerged strongly. There were slow students and fast students, and students requiring remediation. At Lyons, students were sectioned accordingly into "minimum," "regular," and "superior," although these designations were not explicitly revealed to the students. Allen was a strong proponent of such "tracking," a practice that would come under attack in later decades and become immensely politically controversial, especially when race entered the picture.

Allen soon began to display his ambitions for professional advancement. By the late 1950s he had been president of the Men's Mathematics Club of Chicago (see chap. 12) and of the Illinois Council of Teachers of Mathematics and was serving on the board of directors of the National Council of Teachers of Mathematics. He was also promoting specific curricular ideas. He had become impressed with the utility of an idea suggested by Ohio State University mathematics educator Nathan Lazar in the 1930s: the multi-converse concept. The converse of the statement A implies B is the statement B implies A. The truth of the first implication by no means guarantees the truth of the second, and it can be an enlightening student exercise to investigate. To take a simple example, if two numbers are both even, then their product is even. But if a product is even, it is not necessarily true that each of the two factors is even. Lazar and Allen noted that most mathematical theorems are not so simple as A implies B. They might take the form of A, B, C implies D and E. One can then generate several different converses, by interchanging hypotheses and conclusions, and each such converse can be examined for truth or falsity. Allen found that this helped students in precise use of language and in logical thinking, and moreover introduced an element of discovery into the classroom.

Following on from these ideas, Allen thought hard about the value of proving mathematical theorems in high school classrooms. Traditionally, the only place where such proving occurred was in geometry courses, and the method used was the so-called two-column proof. In the left column one listed the sequence of statements in the proof, with each statement paired with a justification in the right column. Allen was aware that professional mathematicians often scoffed at this method for its crudity. But the mathematicians' own conventions of proof, "elegant" though they were sometimes considered, were rife with gaps and mystifying inside knowledge. Even when mathematicians did not explicitly use phrases like "it is easy to see," they were implying exactly that. They offered no good guidance for novices. As an alternative that he found more appropriate, Allen developed what he called a "flow proof," which was essentially a sequence of implications, emphasizing the logical connections at each step. Allen also felt strongly that high school students should encounter proofs not only in geometry, but in algebra as well. This made him amenable to some of the more abstract approaches to algebra in New Math programs, where proof became paramount.

Through the 1950s Allen became increasingly aware of general national rumblings regarding the mathematics curriculum. The University of Illinois Committee on School Mathematics (UICSM), led by Max Beberman, had been a center of new ideas since 1951 and was especially influential in its home state. Allen was never fully enthused by UICSM; it seemed to him to push an overly abstract approach to his subject. But he liked what he heard of the philosophy of the School Mathematics Study Group (SMSG), the largest curricular reform program, founded in the wake of the 1957 launch of the Sputnik satellite by the Soviet Union. Allen was personally much alarmed by the apparent rise of Soviet military capabilities, and he was eager to offer his services.

In the summer of 1958 Allen was tapped to participate in the first SMSG writing session, conducted at Yale University, home institution of SMSG's head, mathematician Edward Begle. The idea was to bring together small groups of mathematicians and schoolteachers to write new curricular material that would more closely reflect modern mathematics as practiced by mathematicians, and at the same time be teachable in the schools. Allen was put in charge creating the eleventh-grade course on algebra and trigonometry. The summer of 1958 was mainly confined to planning, with the bulk of the writing done the following summer, in Boulder, Colorado. These experimental texts, somewhat crudely produced in softcover format, were used in classrooms across the country in the 1960s.

The SMSG textbooks were not intended to be a permanent fixture in the schools, but rather to be used as guidance for a follow-on generation of commercial textbooks. Allen was generally happy with his contribution to the eleventh-grade pilot text, except that he was unable to convince his team to emphasize language and logic as much as he would have preferred. This disappointment eventually led to his writing his own commercial textbook in 1966, *Modern Algebra: A Logical Approach*, in collaboration with Helen Pearson of Purdue University.

Allen fully subscribed to a basic idea of many of the New Math reformers: that school mathematics had fallen substantially behind the leading edge of mathematical knowledge and needed to catch up. He was furthermore convinced that schoolteachers who embraced curricular reform were likely to be rewarded with a rise in status. His own case certainly justified this conviction.

In the early 1960s Frank Allen was thrust into public view regarding mathematics education reform: as an SMSG participant, as an NCTM board member, as the director of a series of NCTM Regional Orientation Conferences to inform school administrators about the new curricular programs, and finally as NCTM president from 1962 to 1964. This was just at the time that voices started to be raised questioning the wisdom of the innovations, so that Allen was right in the front lines of controversy.

One of the loudest critical voices, singled out in Allen's retrospective lament of 1984, was New York University's Morris Kline, an applied mathematician with broad interests. By the early 1960s Kline had published thoughtful books on the history and cultural value of mathematics, well regarded by professionals and the general public alike. His essays on mathematics education, in contrast, were slashing polemics, self-consciously so. The title of a 1956 essay was "Mathematics Texts and Teachers: A Tirade." He entirely agreed with the New Math reformers that school mathematics needed major improvement, but he intensely disliked the proposed new programs. They did not, in his view, sufficiently build abstract concepts on concrete examples, they scanted connections with physical science, and they focused too much on nurturing future mathematicians.

Many opponents found Kline reasonable in private conversation but accused him of always choosing to be as inflammatory as possible when he had an audience. While having no trouble expounding upon the shortcomings of everyone else's curricular proposals, his hints as to his own, presumably brilliant, alternative program never seemed to get fleshed out into a useable form. More than one Kline critic told him to "put up or shut up." SMSG participants, such as Frank Allen, who knew with intimacy the day-to-day difficulties of actually fashioning a new curriculum and trying it out in the classroom, were incensed by Kline's lofty disdain for their hard work. And Kline's claim that schoolteachers involved in New Math programs, such as Allen, were being dictated to by university mathematicians, was infuriating. Allen also considered Kline unscrupulous in tarring SMSG with the flaws of other programs. Looking back, Allen considered Kline's influence as thoroughly detrimental, because "Kline, darn it, was a tremendous speaker." Allen was proud to have held his own the one time he directly debated Kline, at the Peabody Hotel in Memphis.

In 1962 Kline created a special stir by organizing an open letter, signed by 63 prominent mathematicians in addition to himself, laying out his objections to the New Math programs. The letter was published in the *American Mathematical Monthly* and the *Mathematics Teacher*, the official journals of the Mathematical Association of America and the National Council of Teachers of Mathematics, respectively. No specific New Math programs or reformers were named, but SMSG proponents were inclined to view it as an insidious attack. Compared to Kline's earlier diatribes, this open letter was mild, but his opponents sniffed out his involvement. Begle, the SMSG director, queried the distinguished cast of signatories and discovered that they had mainly served to provide a false aura of widespread consensus to the critique; most of them proved to have little if any knowledge of the SMSG experimental texts, and no inclination to actively pitch in to improve them.

Allen continued to be loyal to SMSG, both in the 1960s and later in life. He admitted its imperfections but insisted that its basic aims were valid. The members of the writing teams, he maintained, were knowledgeable and conscientious, but able to laugh at themselves. It was SMSG participants, not the critics, who first offered the joke that SMSG stood for "Some Math, Some Garbage." Ed Begle was delighted by Tom Lehrer's satirical song "New Math." As Allen recalled, "He howled and yelled. I thought he was going to die."

In addition to Morris Kline, Allen was unable to forgive Max Beberman for his contribution to sinking New Math. Beberman, as noted, had been visible on the educational scene even before Sputnik and SMSG. He had come to the University of Illinois in the early 1950s, as he was finishing an education doctorate from Teachers College at Columbia University, and soon teamed up with some Illinois mathematicians to create experimental curricular materials. These were tried out first at the University High School and then, with private and government grants, at a wider selection of schools. The Illinois materials inspired some of the later SMSG materials, although Allen and others considered that Beberman was too fond of obscure symbolic notation and of nitpicky distinctions, such as that between number and numeral. Allen deemed it foolish to belabor the fact that the abstract concept of five was not identical with its various representations, such as 5 or V. But Beberman was a charismatic teacher of both schoolchildren and education graduate students who could, as Allen acknowledged, "charm the birds out of the trees." Thus the Illinois program gained many adherents, and Beberman was widely viewed as an uncompromising leader of New Math.

It was therefore a great shock to many education observers, and especially to Allen, when Beberman in December 1964 expressed public doubts about New Math. Speaking at an NCTM meeting, and reported in the *New York Times*, Beberman expressed his fear that the new programs were being instituted too quickly, with insufficiently prepared teachers. He further urged that more attention be given to showing students the utility of mathematics, not merely its abstract structure. He warned of "charlatans" and "scandal." Allies of Beberman felt he was merely urging a more measured approach, but Allen considered it a fundamental betrayal. The two men were still unreconciled at the time of Beberman's premature death in 1971. In consequence, Allen felt awkward when in 1987 the Illinois Council of Teachers of Mathematics awarded him the Max Beberman Mathematics Educator Award.

Frank Allen had left high school teaching by then, having in 1968 become an associate professor of mathematics at Elmhurst College, in his hometown. He was initially wary of entering the college environment, unarmored with a PhD, but he managed the challenge, taught all the college courses except topology, and ended up as department chairman. He continued to observe school mathematics with great interest, as the New Math reform era sputtered out, given its final death blow in many eyes by the publication in 1973 of Morris Kline's *Why Johnny Can't Add*. The new era, dubbed Back to Basics, dismayed Allen. But he was also distressed by the response of his old organization, the NCTM. Seeking some compromise between New Math and Back to Basics, they adopted what to Allen was a nebulous touting of "problem solving," sacrificing attention to the logical structure of the discipline. This was a large part of the context of his 1984 lament for New Math.

Allen's alienation from the NCTM grew during the 1980s. Leaders of the organization who succeeded him took away different lessons than he had from New Math's demise. In particular, many felt that the NCTM as an organization had been too passive, simply accepting the suggestions of university mathematicians as to what the schools needed. The determination to be more activist and more independent of the mathematicians led to creation of a series of documents, collectively labeled the "NCTM Standards," beginning with the *Curriculum and Evaluation Standards for School Mathematics* of 1989. Supporters of the Standards saw them as encouraging a balanced approach to school mathematics, befitting the new technological landscape in which they lived. With the rise of cheap calculators, less attention to arithmetic algorithms was needed. In contrast, computers and calculators made probability, statistics, and data analysis much more accessible and useful, and consequently they should be given more attention. Students should become skilled in translating between graphical and algebraic representations of problems.

The NCTM Standards became an object of loathing for Allen and like-minded educators. For Allen, the NCTM was neglecting basic skills while promoting wishy-washy problem solving, superfluous applications, and loose logic. He felt that the organization had been captured by what he called the "educational establishment," represented by professors from schools and departments of education, inclined to subordinate the subject matter of mathematics to fads such as cooperative learning and vain attempts to solve the racial, economic, and environmental problems of society through the schools. The NCTM Standards had been primarily written by members of this educational establishment, as opposed to the good old days of New Math, when schoolteachers and mathematicians, properly reverent toward their subject, had been dominant.

The controversy that swirled around the NCTM Standards became known as the Math Wars. Owing to the advance of communications technology since the 1960s, the transmission of the debate exhibited some novel features compared to the New Math era. Articles in newspapers and magazines were now supplemented by websites. (Mercifully, perhaps, social media had not yet arrived.) A site called MathematicallyCorrect.com, founded in 1997, became a favorite platform for polemics against the Standards. Frank Allen posted several documents on this site. Those on the other side posted material on MathematicallySane.com. Mathematically Correct contributors decried materials based on the NCTM Standards as "fuzzy math," "rainforest algebra," and "new new math." Allen was fully in agreement with the first two, but the last epithet made him cringe, as a faithful believer in the original New Math. Many Standards opponents casually conflated the reform proposals of the 1960s and the 1990s. Even such a normally judicious observer as Martin Gardner, the great guru of recreational mathematics, made the historically ill-informed assertion that the NCTM was primarily responsible for New Math.

In late 1999, some historically astute anti-Standards mathematicians made a conscious attempt to duplicate one of the significant events of the

Frank B. Allen, August 12, 1999. Photograph by David L. Roberts

New Math debates, the open letter signed by 64 notables, engineered by Morris Kline in 1962. The immediate issue in 1999 was a set of Standards-friendly mathematics curricula, recently endorsed by US Secretary of Education Richard Riley. A core group of six, led by mathematicians from California, where the debate was especially hot, drafted a letter to Riley, condemning the endorsements as premature and placing emphasis on the assertion that no active research mathematicians had been involved in developing any of the curricula.

More than 200 additional signatories endorsed this letter, the great majority being professors of mathematics. There was also a substantial contingent of physicists, including three Nobel laureates, a smattering of signers from other sciences, and a few general educational controversialists, such as Chester Finn and E. D. Hirsch. Whatever the merits of the letter, the large number of signatures surely reflected the ease of communication now available through email and the Internet. After the letter was drafted, it was posted on the Mathematically Correct website, attracting signatures thereby. Once a substantial list of signers was compiled, it was published in the *Washington Post* on November 18, 1999, taking up an entire page.

Having the letter posted on the web allowed a further communication bonus unavailable in 1962: readers could examine some of the materials referenced in the letter immediately, via hyperlinks. But whether this made the letter signers of 1999 substantially better informed about the issues under discussion than those of 1962 may still be doubted. Despite the impression given by the 200 letter signers, there was in fact no clear-cut consensus among mathematicians on these matters. Hyman Bass, a prominent research mathematician, while agreeing with some of the points made in the Riley letter, feared that it served mostly to poison the political climate in education, especially the lauding of the educational wisdom of mathematicians over other interested parties.

One of the signers of the letter to Secretary Riley, listed near the top by virtue of alphabetical order, was Frank Allen, designated as "Professor of Mathematics Emeritus, Elmhurst College, Former President, National Council of Teachers of Mathematics." Ironically, he was playing a role similar to that played by supporters of his old nemesis of the 1960s, Morris Kline. It was one of the last public displays of Allen's concern about mathematics education. He died in 2003, age ninety-three.

20

Suspicious Minds

John F. Nash Jr.

1994

One's observations of professional mathematicians hardly supports the view that they are the most whole and intact psychological specimens mankind has to offer.

William Barrett

I N 1994 the mathematician John Nash Jr. won a Nobel Prize. It was not the Nobel Prize in mathematics, for there is no such prize. Rather, it was the prize in economics. Nevertheless, the work for which Nash won the prize was indubitably mathematical; it had been the essence of his Princeton dissertation for a mathematics PhD. Nash was legendary among mathematicians, because of both his brilliant early research and his subsequent withdrawal into mental illness. Only a few years prior to the Nobel award, Nash had begun to emerge from decades of schizophrenia, a rare occurrence for a person with his symptoms.

Nash's story resonated with many observers. Journalist Sylvia Nasar was moved to write a biography of Nash, *A Beautiful Mind*, published in 1998. This was followed in 2001 by a "major motion picture," with the same title as the book, starring Russell Crowe as Nash. Playwright David Auburn's *Proof*, which debuted in 2000, featured a Nash-like mathematician whose daughter inherits both his mathematical talent and his mental illness. This too became a movie, in 2006.

Nasar's book on Nash was deeply researched and fluidly written. It ranged widely across Nash's life, celebrating his genius but not shying from describing the pain inflicted on friends and family by his behavior, only some of which could be attributed directly to his mental illness. Nasar sketched a vivid picture of the world of elite university mathematicians in the decades after World War II. She managed to convey something of the difficult content of Nash's mathematics as deftly as could be expected in a popular book.

The movie inspired by Nasar's book, directed by Ron Howard, was more superficial in numerous ways, though satisfying to many viewers as a dramatic narrative. Understandably, the movie depicted Nash's mental illness through visual hallucinations, whereas in real life his hallucinations had all been auditory. The movie also made no attempt to sketch the range of Nash's mathematical achievements. Some critics objected to the simplifying of Nash's complicated sexual life and the minimization of his arrogance, snobbery, and outright meanness. A. O. Scott of the *New York Times* panned the movie's depiction of the activities of Nash and other mathematicians during the Cold War. Nash was not the code-cracker of the movie, and American mathematicians, according to Scott, were far from presenting a united front in the struggle against the Soviet Union.

Scott was nevertheless appreciative of Crowe's performance, especially the actor's subdued portrayal of the travails of schizophrenia. Mental health advocates were generally pleased with the movie in this regard. Lynne M. Butler, a mathematician with a schizophrenic brother, applauded the film for treating the affliction "seriously and sensitively."

Many nonmathematicians are willing, sometimes eager, to see most mathematicians as social misfits. Sylvia Nasar could not resist commenting whenever she encountered social skills in a mathematician or a theoretical physicist. She described theoretical physicist Freeman Dyson as having "an acute interest in people, unusual for someone in his profession." And mathematician Harold Kuhn was a "sophisticated man," free of "the mathematical personality." At times Nasar depicted the mathematical research world as fundamentally characterized by arrogance, immaturity, and insecurity, with Nash merely being an extreme example.

Nash was first introduced to this world in the early 1940s when he was thirteen or fourteen, when he read E. T. Bell's *Men of Mathematics*. As recounted in chapter 11 of the present work, Bell's book presented the field as dominated by an elite of intrepid great men, solving the great problems.

Although Bell did mention instances of collaboration, it is easy to extract from his book an impression that mathematics is mostly done in isolation, with most of the collaboration that occurs being done diachronically, across generations, as in Newton's famous statement that he had "stood on the shoulders of giants." Nash did not work in isolation in the sense of never talking to other mathematicians; at the height of his powers he would frequently pick other's brains and run ideas by them. But he always made it clear that he was seeking sole credit for his original ideas, and he was mightily disappointed when he was scooped. He wrote some joint papers in his early game theory period, but later displayed no interest in seeking out colleagues as coauthors.

═══

Nash was born in West Virginia in 1928 and grew up in comfortable circumstances. Both parents were college graduates. When Nash entered the Carnegie Institute of Technology in Pittsburgh, he initially seemed headed toward engineering, his father's field. But as Nasar pointed out, Carnegie after World War II received an infusion of talented young professors of more theoretical bent. So when Nash became annoyed with his engineering and chemistry courses, he was welcomed into mathematics by professors well able to give him a solid background for appreciating the latest developments in the subject.

During one of the years that Nash was an undergraduate at Carnegie Tech, Joe Diaz (see chap. 18) was an assistant professor. Another professor was Alexander Weinstein, whom Nasar noted as being sought by Albert Einstein as a research associate in the early 1930s. Diaz and Weinstein soon began a long-term mathematical partnership. Unlike Nash, Diaz left Carnegie convinced of the centrality of collaboration to mathematical progress.

Before he finished at Carnegie, Nash applied to only the best graduate schools in the country. He would have preferred Harvard, whose mystique always dazzled him, but he ended up at Princeton, which offered more financial support. Princeton mathematics had been rapidly rising in promi-

John Forbes Nash Jr., 1950. Graduate Alumni Records, Box 119, Princeton University Archives, Department of Rare Books and Special Collections, Princeton University Library

nence over the preceding two decades. In addition to the excellent university department, there was also the nearby Institute for Advanced Study (IAS), with a collection of titanic scholars in mathematics and theoretical physics. Said scholars, including Albert Einstein, had no required duties whatsoever but were sometimes agreeable to consultation, and they occasionally gave lectures. The IAS was the apex of the ivory tower. Princeton was thus a wonderful place to learn mathematics, but its aristocratic isolation would further encourage Nash in his elitist direction.

Many mathematicians have reveled in their subject for aesthetic reasons. Serge Lang, who was a graduate student at Princeton at the same time as Nash, when asked what pure mathematics was good for, replied, "It's good to give chills in the spine to a certain number of people, me included. I don't know what else it is good for, and I don't care." For Nash such aesthetic pleasures seem to have been secondary to an intense desire to be recognized for his superior ability. He wanted to be first to solve problems, and he wanted to solve problems that were deemed important by other mathematicians. His thirst for recognition surfaced already at Carnegie

Tech when he participated in the Putnam Competition. This challenging nationwide test for undergraduates, run by the Mathematical Association of America, has since the 1930s been considered an excellent measure of superior mathematical talent. The top five finishers receive special recognition by being named Putnam Fellows. Nash's failure to become a Putnam Fellow rankled so much that he alluded to it in his Nobel Prize acceptance speech 45 years later.

Appropriately, Nash's Princeton dissertation was about competition, being a contribution to the then-new mathematical subfield of game theory. Game theory offers a quantitative analysis of contests between two or more "players," each of whom makes a series of "moves" and at the end receives a "payoff" that is either positive, negative, or zero. Familiar recreational board and card games can be illuminated by the theory, but it was early recognized that there were potential applications to more momentous situations in politics and economics. The foundational book in the field was *Theory of Games and Economic Behavior*, published in 1944 by mathematician John von Neumann (of the Institute for Advanced Study) and economist Oskar Morgenstern. But this book was largely confined to the simplest case of two-person, zero-sum games, where one player's winnings exactly match the other player's losses.

Nash's dissertation was a decisive advance, showing convincingly how to analyze multiplayer, non-zero-sum games, without artificial assumptions of cooperation between players. This made the theory substantially more relevant to economics and would in time lead to Nash's Nobel Prize. Nash translated the game theory context into geometry, so that he could use a powerful tool of twentieth-century mathematics: the fixed-point property of certain geometric objects. When such an object is transformed into itself in a continuous manner, at least one point must remain fixed. (For example, a circular disk rotating around its center.) The existence of this fixed point implied that there must be an equilibrium point for the game: a set of strategies for all the players where no player benefits by changing strategy.

As noted in chapter 18, World War II and the subsequent Cold War confrontation with the Soviet Union drew many American mathematicians into relationships with the military. Early in the Cold War period, game theory became one of a suite of mathematical techniques, including linear programming and aspects of probability and statistics, used by the military for resource allocation and decision making. Collectively these techniques became known as operations research, for the focus here was not on designing and building weaponry and other machines of war (plenty of mathematics was well recognized to be useful in these activities), but in the optimal operation of such machines to achieve a desired objective. The RAND Corporation of Santa Monica, California, a private firm with military contracts, became a leading center for operations research and for game theory in particular. RAND was a "think tank," in the parlance of the time.

Nash's game theory work led to his being hired by RAND as a consultant for several summers in the early 1950s. This experience gave Nash some handy extra income and some stimulating contacts with other mathematicians, but it ended abruptly in 1954 when he was arrested for indecent exposure. This was a time when fears about communist subversion had been stoked by demagogues such as Senator Joseph McCarthy, and homosexual behavior by persons in sensitive positions was considered an automatic security threat, because of an alleged susceptibility to blackmail. RAND was able to control publicity surrounding Nash's arrest, so that his academic career was not directly affected, but they immediately dismissed him. Nash's usefulness to the company had greatly diminished anyway, as by then he had largely lost his interest in game theory, preferring more abstract realms of mathematics.

The mathematicians who know Nash's accomplishments best are generally of the opinion that his pure mathematics work of the 1950s decisively surpasses his game theory breakthrough, Nobel Prize notwithstanding. Nash spent that decade with faculty positions at the Massachusetts Institute of Technology and New York University, and with one fellowship year back in Princeton at the IAS. During this time he dazzled his colleagues by solving difficult problems in geometry and analysis. He demonstrated that certain convoluted geometric objects, defined in opaque ways, can be "embedded" in more familiar Euclidean spaces without changing their fundamental properties. He also proved striking theorems characterizing the solutions of partial differential equations. He used the results of other mathematicians, and sometimes solicited their help, but the final constructions were so astonishingly unforeseen that no one gainsaid his originality.

Some of Nash's greatest mathematical contemporaries have been moved to use the word "genius" to describe him. This word can be confounding, even dispiriting, to mathematics educators, who need faith in the fundamental malleability of human beings, believing that anyone with unimpeded mental apparatus can, with suitable instruction, be brought up to whatever level of mathematical knowledge and accomplishment is needed in a particular environment. But a person like Nash brings this all into question, suggesting that some people are born with talents so above the ordinary that education is essentially irrelevant.

Nash's work on partial differential equations also resulted in a psychological bruise related to his competitive streak, and there has been speculation that this may have triggered the first uncontrolled exhibition of his schizophrenia. Nash discovered that a key result he was preparing for publication had already been done by the Italian mathematician Ennio De Giorgi. Here again Nash's regret was such that he mentioned it in his Nobel Prize address. Although there was overlap between the Nash and De Giorgi results, their approaches were not identical, and most qualified observers feel that neither detracts from the achievement of the other. Nevertheless, Nash blamed the coincidence on his failure to win an award he especially coveted, the Fields Medal, named after J. C. Fields, the Canadian mathematician who endowed it in the 1930s. Because there is no Nobel Prize in mathematics (juicy stories attributing this oversight to Alfred Nobel's sexual jealousy of a Swedish mathematician circulated for many years but have been thoroughly debunked), the Fields Medal is generally considered the most prestigious in the discipline, although its monetary value is surpassed by other awards. Although there can be as many as four Fields Medal recipients at the same time, there are restrictions that increase its rarity: it is awarded only every four years, and awardees must be under age forty. When Nash failed to win in 1958, at age thirty, many were confident that he would surely win in 1962 or 1966. Unhappily, it was in the 1960s that Nash's mentally disturbed behavior manifested itself most severely.

In 1958, the year of Nash's disappointment with the Fields Medal, he was featured in a popular article on the state of modern mathematics in *Fortune* magazine. He was one of only four mathematicians distinguished with a photograph. The accompanying text accurately touted him as "eager to tackle the most difficult problems" and as "one of the few young mathematicians who have done important work in both pure and applied mathematics." The article, by journalist George A. W. Boehm, rapidly sketched the landscape of current mathematics, including game theory, number theory, logic, and differential equations. The advent of computers was noted. Boehm asserted that the large increase in demand for mathematicians in government, industry, and academia was creating a "crisis in education," leading to proposals to revamp secondary school mathematics instruction. This movement came to be known as "New Math," although Boehm never used that term. He mentioned two leaders in this effort by name: Edward Begle of Yale and Max Beberman at the University of Illinois (see chap. 19).

That Nash would be referenced along with a discussion of educational issues was incongruous. Although there are reports of his inspiring some students, he was commonly judged to be a poor teacher, and he never expressed any interest in the problems of school education at all.

The year 1958 would prove to be the high point of Nash's early career. He was married, with a child on the way. His faculty appointment at MIT was coming to an end, but job offers elsewhere were coming in. (But not from Harvard, his heart's desire.) He had not won the prizes he had hoped for, but he was well regarded as one of the brightest stars of his mathematical generation. In early 1959, however, he began to alarm friends and family with his talk and behavior. He gave incoherent lectures. He claimed that secret messages were being directed to him personally, hidden on the front page of the *New York Times*. He tried to use the MIT departmental mail to send strange letters to foreign ambassadors. He turned down a professorship at the University of Chicago because he claimed to have received a prior offer to become emperor of Antarctica.

The next decade was chaotic. Nash's wife, at times physically fearful of him, had him involuntarily committed to mental institutions on several occasions. He received insulin treatments and Thorazine, with unclear results. Early on he was in the McLean Hospital outside Boston, in Bowditch Hall, named for a philanthropist who had taken interest in Boston's medical facilities in the nineteenth century, Nathaniel Ingersoll Bowditch (one of the sons of Nathaniel Bowditch, the navigator of chap. 1). This high-class facility, unaffordable in the long run for Nash's family, featured some celebrated patients. Here Nash encountered poet Robert Lowell, who stayed at McLean repeatedly to treat his manic depression and wrote a poem with the line "This is the way day breaks in Bowditch Hall at McLean's."

When not institutionalized, Nash lived in Boston, Princeton, and Roanoke, Virginia, where his sister lived. He also took trips to Europe, where he tried to renounce his American citizenship. He spoke of leading vast projects to bring about world peace. During brief periods of lucidity, he managed to publish a couple more significant research papers on partial differential equations. But more often he was obsessed with numerological patterns, hinting that he could discern momentous political implications therein. Sometimes he fastened on specific controversial issues, such as the plight of Palestinian refugees.

But the greatest political controversy of the 1960s, the Vietnam War, seems not to have especially engaged his attention. University mathematicians across the country were not insulated from the effects of that war. In 1970 a bomb was exploded at the Mathematical Research Center of the University of Wisconsin in Madison, killing one and injuring three. In 1971 Serge Lang, Nash's old student colleague at Princeton, resigned from Columbia to protest the university's treatment of antiwar protesters. At Princeton there were repeated protests targeting the relationship between the university and the Institute for Defense Analyses, a military think tank whose Princeton branch was overseen by a Princeton mathematics PhD, Lee Neuwirth. Through part of this period Nash was absent from Princeton, spending time instead in Roanoke. For the rest of the period he was preoccupied with his private demons.

Nash's wife divorced him, but beginning in 1970 she allowed him to live with her in Princeton. Her idea, ratified by subsequent developments, was that in these familiar surroundings he might be able to attain some mental equilibrium. He was given some limited privileges at the university, and although sometimes disturbing those unused to his odd mannerisms, he was able to come and go with only minor upset to the regular campus life. Slowly, over two decades, there was a tapering off of the outward symptoms of his illness. More and more, when he talked mathematics, he displayed sense, even acute perception, rather than numerological fantasy. Thus was the scene set for the triumph of the Nobel Prize of 1994.

The last segment of Nash's life featured additional successes but ended with unhappy suddenness. He and his wife remarried. Colleagues found him less arrogant and less mean. He launched some new research projects, in collaboration with younger mathematicians. And he was rewarded with yet another accolade, the 2015 Abel Prize. This honor, presented by the Norwegian Academy of Sciences, is named after the famed nineteenth-century Norwegian mathematician Niels Henrik Abel. First awarded in 2003, it has rapidly attained a reputation among mathematicians nearly equal to that of the Fields Medal and is considerably more financially lucrative. There is no age restriction, and so far it has recognized senior contributors. On returning from the Abel Prize ceremony in Norway, John Nash and his wife were killed in a traffic accident while riding in a taxi on the New Jersey Turnpike.

Conclusion

It is getting easier to make the calculations, what with the decimal notation, logarithms, slide rules, desk computing machines, and the great computing machines which are getting larger in capacity as they get smaller in size due to miniaturization. The question is: Are we getting smarter as fast as our tools are improving or are we becoming more likely to cut ourselves intellectually to bits with the increasing sharpness of the tools?

E. B. Wilson

BY ALMOST ANY MEASURE, mathematics is far more embedded in American life in the opening decades of the twenty-first century than it had been at the end of the eighteenth. It is pervasive throughout the education system. It is also pervasive, though less clearly, in the enormous array of technological gadgets that would have astonished Americans living in 1800. Mathematics appears in popular culture as well. The book and movie versions of *A Beautiful Mind* (discussed in chap. 20) found a wide audience at the turn of the twenty-first century. More recently, *Hidden Figures*, which tells the stories of African American women mathematicians working in the early US space program, has likewise been enthusiastically received in both formats. An array of books for popular audiences on mathematics and mathematicians have garnered more modest successes. No such accounts were available during the early years of the republic.

A country that once had nine colleges that gave modest attention to mathematics now has hundreds harboring research mathematicians. According to the American Mathematical Society, in 2018 there were 244 institutions that award the PhD in mathematics, 177 more where a master's is the highest mathematics degree, and a further 1,017 colleges where one may earn a bachelor's degree in the subject. Then there are the community colleges, all of which teach some mathematics and many of which require it for virtually any degree. The universities, colleges, and community colleges are fed by a decentralized system of elementary and secondary schools, public and private, across the land. One would be hard pressed to find any school in which there is no attention to mathematics, whereas an eighteenth-century "grammar school" might literally have been devoted to reading and writing only.

The life stories in the preceding chapters indicate an unmistakable tendency for involvement with mathematics to entail increasing levels of formal education. A figure such as Nathaniel Bowditch, who in the early nineteenth century made contributions to mathematics despite meager schooling, was already wildly improbable by the end of that century. By then it was likewise unthinkable to write an arithmetic textbook without any college education, as Catherine Beecher did in the 1830s. Of the individuals discussed in this book, the only ones with no college all reached maturity before 1840. Those who lived mainly in the twentieth century not only went to college but also earned a master's or doctoral degree.

Admittedly, this book exaggerates the speed of these transformations. The PhD did not emerge so rapidly in the nineteenth century, nor become so prevalent in the first half of the twentieth, as one might deduce purely from reading this book. Even in the twenty-first century one can teach school mathematics without a master's degree.

It is clear from the lives featured in this book that investment in education has not been evenly distributed in time or place. Opportunities for learning mathematics, and for using mathematics in a career, have frequently depended on the fortunate circumstances of a person's local environment, although individual resourcefulness can overcome many obstacles, as in the case of Abraham Lincoln. Discrimination on the basis of race, ethnicity, and gender plays a role in some of these stories, but the full effects of these practices, long omnipresent and still obstinately lingering today, cannot be assessed from such a small selection. Nor can the extent of ethnic diversity among mathematical practitioners be accurately judged from the limited offerings here. Only glancing references are made, for example, to the substantial contributions of Americans with ethnic roots in Asia.

Yet the widespread presence of mathematics in American lives has not led to uniform appreciation of its value, or agreement on how much every American should know about the subject. The large amount of attention given to mathematics in schools has led to a psychological affliction sometimes called "math anxiety," a condition not readily discernible in the nineteenth century. And most people who have publicly announced some knowledge of mathematics have encountered friends, relatives, or random strangers who respond by admitting their ignorance or distaste, sometimes proudly. But some mathematicians are perfectly happy to be viewed as engaged with lofty, incomprehensible matters.

From this book it can be observed that disagreements on teaching mathematics emerged already in the early twentieth century, when Americans first began to think that there was such a thing as a national education system. The "New Math" of the 1950s and 1960s was an especially lively time of controversy. Although this episode is often considered to be an ephemeral failure, my own view, perhaps evident in several of the later chapters, is that it had lasting positive effects that should not be casually discounted. Disputes surrounding the content and methods of mathematics instruction have continued into the twenty-first century without much resolution, as many readers will be aware.

An interesting fact of which I am well convinced, although I have never attempted to assemble any data to support it, is that the past two centuries have witnessed a general convergence of educational institutions, at all levels. Secondary schools today are more alike in form than any two secondary schools one might have picked in the early nineteenth century. West Point and Harvard are far more similar now than they were in 1815. This convergence has occurred despite the lack of national control of education. Reasons for it include state regulations, regional accreditation agencies, and the consolidation of the textbook industry. Also, a few prominent institutions have inspired emulation. There now exist more competitive pressures and more data for assessments. No one in 1800 was concerned to collect numerical measurements to compare the schools of one state with those of another. Today, there are many interested in such numbers. At the college level the employment churn surely plays a role, as faculty members move frequently between schools, carrying attitudes and methodologies.

To say that schools are increasingly alike in form does not mean that they are increasingly alike in resources or results. Now that almost all secondary schools are essentially pursuing the same goals, with the same basic structure, inequities are more evident today than 200 years ago. Although a graduate of the most impoverished high school should in theory be able to emerge some years later with a PhD and a fellowship at the Institute for Advanced Study, the odds against such achievement are enormous.

Some believe that even the best schools are of minor importance when it comes to producing a true grasp of mathematics. Educators might ponder the view of one of the greatest American mathematicians of recent years, the late William Thurston: "It is very hard to get a sense of the depth, liveliness, power, and breadth of mathematics from any ordinary experience with mathematics in school. I believe that most students who really master the subject, and eventually become scientists, mathematicians, computer programmers, etc., are those who have some other channel for learning mathematics outside the classroom."

E. T. Bell's *Men of Mathematics*, whose shadow to some extent falls over the present book, portrays the subject as largely driven by a few exceptional individuals. I do not deny the existence of such people, from Nathaniel Bowditch in chapter 1 to John Nash in chapter 20, with the towering figure of Willard Gibbs in between. But I have chosen to write also about less lofty figures, and to frequently emphasize mathematics as a social activity. I note the importance of collaboration and outreach, from Sylvanus Thayer's teaching incubator at West Point to Joe Diaz's jointly authored research papers.

Certain institutions appear repeatedly in these stories, and in many cases this is a fair representation of their overall influence on the mathematical life of the nation. The US Military Academy at West Point was indubitably significant for training engineers and teachers in the nine-teenth century, and the colonial colleges, especially Harvard, have played the substantial role suggested by their frequent mention. Of newer univer-sities, Johns Hopkins and the University of Chicago receive attention in this book somewhat in accordance with their significance. But there are many important universities mentioned only in passing or not at all, and the same is true of mathematical practitioners. Readers of this book will perhaps gain some inkling of the stature of Benjamin Peirce, but I have said nothing about earlier nineteenth-century figures such as Robert Adrain, Jeremiah Day, or John Farrar. Many notable late nineteenth- and twentieth-century mathematicians go unmentioned.

But a more significant deficiency of the book, in my view, is its slight attention to the use of mathematics in industry in the twentieth century. Grace Hopper is the only representative. In fact, after World War II, in ad-dition to the computer business, there were numerous mathematicians em-ployed in such industries as automobiles, petroleum, and telecommunica-tions, most notably at the celebrated Bell Telephone Laboratories.

Computers, discussed a bit in chapters 10 and 15, are the most obvious example of the extent to which mathematics has become embodied in tech-nology over the past 200 years. This development has been proclaimed as a reason that students should study mathematics, and certainly some people will need to know mathematics in order to continue to make sophis-ticated computer applications, but the implications for the education of the great majority of people are less clear. The rise of the automobile in the twentieth century did not seem to require widespread understanding of the technicalities of the internal combustion engine. In many ways, comput-ers negate the need for mathematical knowledge. This may eventually en-tail dramatic changes in mathematics education. One mathematics profes-sor sounded the alarm in 1997: "How long will it be before we have a 200 Mhz computer with 32 Mb of RAM that slips into a shirt pocket? When that happens many disciplines that presently require their students to take mathematics courses may begin to reconsider such a requirement. If what they want is that their students learn how to execute certain algorithms, I don't see why they should keep sending them to us." Since compact devices far surpassing the above specifications are now available, the pressures on mathematics education would appear to be all the greater.

Computers have begun placing pressure on even the highest levels of mathematics research and mathematics education, as alluded in chapter 10, with the mention of the "death of proof." One of the most ambitious efforts in this direction has been made by the British American computer scientist Stephen Wolfram. He believes that "the current predominance of partial differential equations is in many respects a historical accident." The future of mathematics, according to Wolfram, belongs to computer modeling, specifically using so-called cellular automata. He set out his immodest views in a massive book in 2002: "So what this means is that in the future, when the ideas and methods of this book have successfully been absorbed, the field of mathematics as it exists today will come to be seen as a small and surprisingly uncharacteristic sample of what is actually possible."

As someone who grew up in the United States in the second half of the twentieth century, I have a personal attachment to many aspects of mathematics as practiced at that time. But the lives recounted in this book do not suggest a static future. As a historian, I see no reason to expect that the customs of mathematical theorists, technologists, educators, and popularizers will remain comfortably familiar for long.

Acknowledgments

The idea for this book came from my original editor at Johns Hopkins University Press, Vincent J. Burke. Let me state that more strongly: I would never have conceived of this book had not Vince suggested it. Thus I am indebted to him for the many pleasures I experienced during the research and writing. Once I became fully invested, the project suited me well. In making the book my own, I have inevitably strayed somewhat from his original conception. In particular, the chapters do not focus tightly on a single decade. Each chapter does begin with an episode in a different decade but then ranges widely, before and after. I also thank Vince for pushing me to make the book more accessible to the general reader.

My dear friend Tomás Kalmar has been a source of general intellectual sustenance for many years, particularly while I was writing this book. He is well aware of my areas of ignorance but more often inclined, with his generous spirit, to buck me up with flattery of my literary skills: "When you write 'grappled,' you don't mean 'wrestled'!" His many stories of his experiences at Harvard ensure that I do not overly venerate that institution. Tomás has discussed this book with me on many occasions from its first inception and has read and commented in detail on several of its chapters. I have dedicated this book to him.

I have gained considerably from the advice of the University Town Center Writers' Group at Prince George's Community College (PGCC), especially Mary Dutterer and Dennis Huffman. I gratefully acknowledge the assistance of PGCC librarians Angela Abrams and Candice Floyd, and I am especially appreciative of Lori LaFontaine for expeditiously dealing with my interlibrary loan requests.

Peggy Kidwell, of the Smithsonian Institution's National Museum of American History, has long been an encouraging historical mentor. For this book in particular she pointed me to helpful sources for the life of Grace Murray Hopper.

An anonymous reviewer of my manuscript brightened my day with encouraging words, while also making valuable suggestions for improvement. Following Vince Burke's retirement from Johns Hopkins University Press, the editorial process was smoothly transferred to senior acquisitions editor Tiffany Gasbarrini, ably assisted by Esther Rodriguez. Thanks also to production editor Kimberly Johnson and managing editor Juliana McCarthy. Freelance copy editor Ashleigh McKown tightened up my prose cheerfully and efficiently.

My wife, Jenny Scott, has provided the environment I needed for writing, and she has patiently tolerated me during the process. I won't claim that all instances of my distracted behavior during the writing of this book were due to my musing on how to improve a paragraph, a sentence, or a chapter title, but some of them have been.

Selected Bibliography

Here I describe the sources for all direct quotations, unless sufficiently identified in the text. Other significant sources for biographical information are also mentioned, if they are not previously mentioned with regard to the quotations or identified in the text.

Introduction

The epigraph is from Benjamin Silliman, "Introductory Remarks," *American Journal of Science and Arts* 1 (1819): 7. The quote from Abigail Adams is from *Letters* (New York: Library of America, 2016), 420. The quote from John Adams is from *Revolutionary Writings 1775-1783* (New York: Library of America, 2011), 564. The quote from Thomas Jefferson is from *Writings* (New York: Library of America, 1984), 1064. Spelling and capitalization are as they appear in the originals.

Chapter 1. A Practical Navigator

The epigraph is from Baron de Zach, *Correspondence Astronomique, Géographique, Hydrographique, et Statistique* (Génes: Luc Carniglia, 1824), 224–25 (my translation from Bowditch's French). Zach's assessment of Bowditch as America's only great geometer is from this same source, page 223. The diary entry is from *The Diary of William Bentley*, vol. 3 (Salem, MA: Essex Institute, 1911), 289. Bowditch's views on slavery are cited in Nathaniel Ingersoll Bowditch, *Memoir of Nathaniel Bowditch* (Boston: Little, Brown, 1840), 116. The description of the origin of books for the Salem Athenaeum is from this same source, page 23. Bowditch on longitude is quoted from *The New American Practical Navigator* (New York: Edmund M. Blunt, 1821), 167. Nathaniel Hawthorne described the decline of Salem in *The Scarlet Letter*, in *Novels* (New York: Library of America, 1983), 123, 144. A significant source for this chapter was Tamara Plakins Thornton, *Nathaniel Bowditch and the Power of Numbers* (Chapel Hill: University of North Carolina Press, 2016).

Chapter 2. Hudson River School

The epigraph is from Sidney Forman, *West Point: A History of the United States Military Academy* (New York: Columbia University Press, 1950), 139. Uncle Toby is quoted from Laurence Sterne, *Tristram Shandy* (Baltimore: Penguin, 1967), 129–30. Dennis Mahan's words on fortification are from an edition pirated by the Confederacy during the Civil War: D. H. Mahan, *Summary of the Course of Permanent Fortification and of the Attack and Defence of Permanent Works for the Use of the Cadets of the U. S. Military Academy* (Richmond: West & Johnston, 1863), 222. His remark on "feet and inches" is from D. H. Mahan, *A Treatise on Field Fortification* (New York: John Wiley, 1856), vi. The statistics on West Point graduates come from "U.S. Colleges and Universities," Fred Rickey's Home Page, accessed March 4, 2019, http://fredrickey.info/dms/OldestSchools.html. Two significant sources for this chapter were Theodore J. Crackel, *West Point: A Bicentennial History* (Lawrence: University of Kansas Press, 2002), and Joe Albree, David C. Arney, and V. Frederick Rickey, *A Station Favorable to the Pursuits of Science: Primary Materials in the History of Mathematics at the United States Military Academy* (Providence, RI: American Mathematical Society, 2000).

Chapter 3. Political Arithmetic

The epigraph is from a speech given by Lincoln in Cincinnati, Ohio, on September 17, 1859. He was arguing against the ideas of Stephen Douglas, against whom Lincoln had waged an unsuccessful campaign in 1858 for a US Senate seat from Illinois. Both men would run for president in 1860. The Cincinnati speech is printed in Abraham Lincoln, *Speeches and Writings 1859–1865* (New York: Library of America, 1989), 68. Lincoln on his educational environment is from the same volume, page 107. Page 726 cites an Indiana senator calling the Declaration of Independence "a self-evident lie." The excerpts from Lincoln's cyphering book are from Nerida F. Ellerton and M. A. (Ken) Clements, *Abraham Lincoln's Cyphering Book and Ten Other Extraordinary Cyphering Books* (New York: Springer, 2014), 156–58. Spelling and missing letters as in original. Ellerton and Clements (page 132) are also the source for the claim that Lincoln's time in school was "above the average." Herndon on Lincoln's ambition is quoted in Richard Hofstadter, "Abraham Lincoln and the Self-Made Myth," in *The American Political Tradition* (New York: Vintage, 1948), 93. Frederick Douglass recounted his master's views on education in *The Narrative of the Life of Frederick Douglass*, in *Autobiographies* (New York: Library of America, 1994), 37. Emphasis original. Lincoln's claim to have been a slave is cited in Allen C. Guelzo, *Abraham Lincoln: Redeemer President* (Grand Rapids. MI: Eerdmans, 1999), 121. Herndon on the snoring circuit riders is from William H. Herndon and Jesse William Weik, *Herndon's Lincoln*, vol. 2 (Chicago: Belford Clarke, 1890), 308–9. All Euclid quotes are from John Playfair, *Elements of Geometry: Containing the First Six Books of Euclid, with a Supplement on the Quadrature of the Circle, and the Geometry of Solids: To Which Are added Elements of Plane and Spherical Trigonometry* (New York: W. E. Dean, 1843), 5, 6, 8, 9, 11, 40, 284. The final Lincoln quote, on discoveries and inventions, can be found online in the *Collected Works of Abraham Lincoln*, volume 2, page 442, accessed August 18, 2018, https://quod.lib.umich.edu/l/lincoln/lincoln2/1:483?rgn=div1;submit=Go;subview =detail;type=simple;view=fulltext;q1=discoveries.

Chapter 4. Textbook Messages

The epigraph is from 1823, as quoted in Kathryn Kish Sklar, *Catherine Beecher: A Study in American Domesticity* (New York: Norton, 1976), 52. Lyman Beecher calls Cincinnati the "London of the west" on page 102. Cincinnati is called "porkopolis" in Daniel Aaron, *Cincinnati: Queen City of the West 1819–1838* (Columbus: Ohio State University Press, 1992), 21. Catherine Beecher called arithmetic "difficult and uninteresting" in her *Educational Reminiscences and Suggestions* (New York: J. B. Ford, 1874), 29. On page 28 she labeled an arithmetic textbook by Nathan Daboll as "mystical." Catherine's remarks on "management of infants" and "frivolous" discussion of equality of the sexes are from her *Treatise on Domestic Economy for the Use of Young Ladies at Home and at School* (Boston: T. H. Webb, 1843), 7, 155. Harriet speaks of an "intermediate" position on slavery in Charles Edward Stowe, *Life of Harriet Beecher Stowe, Compiled from Her Letters and Journals* (London: Sampson Low, Marston, Searle & Rivington, 1889), 88. The problem on dating the Christian era is from Joseph Ray, *Arithmetic Part 3rd*, 16th ed. (Cincinnati: Winthrop B. Smith, 1842), 23. The steamboat problem and the sheep problem are from Joseph Ray, *Algebra, Part Second* (Cincinnati: Sargent, Wilson & Hinkle, 1852), 112, 203.

Chapter 5. Learning to Count

The epigraph is from an interview of Edwin Wilson by R. Bruce Lindsay and W. J. King on June 3, 1964, Niels Bohr Library and Archives, American Institute of Physics, College Park, Maryland. This is also the source for the final quotation in the chapter, that "everyone should do what he wants to do." The descriptions of the career and views of the senior Gibbs are from George P. Fisher, "Rev. Professor Fisher's Discourse Commemorative of Professor Josiah W. Gibbs, LL.D.," *New Englander* 19, no. 75 (July 1861): 605–20. D. J. Struik expresses qualms about the social conscience of the younger Gibbs in his review of Lynde Wheeler's biography of Gibbs, in *Isis* 42 (October 1951): 259. Gibbs's publishing venue is called "obscure" in Keith J. Laidler, "Van 't Hoff and the Scientific Imagination," in *Culture of Chemistry*, edited by B. Hargittai and I. Hargittai (New York: Springer, 2015), 302; I. Bernard Cohen in his review of Wheeler's biography in *Scientific American* (November 1951): 74; and the entry on Gibbs by Martin J. Klein in the *Dictionary of Scientific Biography*, edited by Charles Coulston Gillispie (New York: Scribner, 1972), 5:389. Henri Poincaré's assessment of Gibbs's *Statistical Mechanics* is from *The Value of Science*, trans. George Bruce Halsted (New York: Dover, 1958), 97, which was originally published in 1913. The French original is "un livre trop peu lu parce qu'il est un peu difficile à lire." See Henri Poincaré, *La Valeur de la Science* (Flammarion, 1970), 130, which was originally published in 1905. Robert Bruce Lindsay attributes the "little book" aphorism to Poincaré in his *Introduction to Physical*

Statistics (New York: Dover, 1968; originally published 1941), 103. Lindsay was a student of Wilson's, and in the Wilson interview he clearly suggests that Wilson passed on the aphorism as a translation of Poincaré. The present writer, Lindsay's grandson, heard this spoken aphorism from him on more than one occasion. Muriel Rukeyser's poetic version of Gibbs is in *A Turning Wind* (New York: Viking Press, 1939). Gibbs's assertion that "mathematics is a language" is from her biography, *Willard Gibbs* (Garden City, NY: Doubleday, Doran, 1942), 280. Gibbs is described as a "happy man" in Lynde Phelps Wheeler, *Josiah Willard Gibbs: The History of a Great Mind* (New Haven, CT: Yale University Press, 1951), 176. The quotes from Thomas Pynchon are from his *Against the Day* (New York: Viking, 2006), 156, 158, 319, 526, 548. This chapter also benefited from Bart Kahr, "Gibbs and Amistad," in *Culture of Chemistry*, edited by B. Hargittai and I. Hargittai (New York: Springer, 2015), 249–56.

Chapter 6. Naval Reserve

The epigraph is from J. W. N. Sullivan, *The Limitations of Science* (New York: Mentor Books, 1949), 164. The son of Admiral Charles H. Davis describes his father's "early experience in fighting and danger and adventure" in Charles H. Davis, *Life of Charles Henry Davis* (Boston: Houghton Mifflin, 1899), 45. The son describes Cambridge "intellectual life" before the Civil War in Charles H. Davis, *Biographical Memoir of Charles Henry Davis 1807–1877* (Washington, DC: National Academy of Sciences, 1896), 31.

Chapter 7. General Principles

The epigraph is from D. H. Hill, *Elements of Algebra* (Philadelphia: J. B. Lippincott, 1857), 13. Hill's account of the South Mountain incident is in Robert Underwood Johnson and Clarence Clough Buel, eds., *Battles and Leaders of the Civil War* (New York: Century, 1884), 2: 562. His assessment of Ulysses S. Grant comes from volume 3, page 638. Hill's prediction of "waves of oblivion" is from volume 2, page 581. Grant's comments on the Mexican War are from Ulysses S. Grant, *Personal Memoirs* (New York: Library of America, 1990), 41. Hill cites mathematics as the "essential pre-requisite" in D. H. Hill, "Education," *The Land We Love* 1 (May 1866): 4. Page 11 of this same article contains his remark on the Southern college president. A significant source for this chapter was Hal Bridges, *Lee's Maverick General: Daniel Harvey Hill* (Lincoln: University of Nebraska Press, 1991).

Chapter 8. Fellow Worker

The epigraph is from a letter of J. J. Sylvester to Daniel C. Gilman, April 2, 1878, quoted in Karen Hunger Parshall and David E. Rowe, *The American Mathematical Research Community, 1876–1900: J. J. Sylvester, Felix Klein, and E. H. Moore* (Providence, RI: American Mathematical Society, 1994), 84. The comments on the Boston Public Library are from Christine Ladd, "Crelle's Journal," *The Analyst* 2

(March 1875): 51–52. The typesetting predicament is noted in Christine Ladd, "Quaternions," *The Analyst* 4 (November 1877): 174. Ladd-Franklin's recollections of Sylvester and Peirce, and her closing comment on the "joy of the intellectual life" at Johns Hopkins, are recorded in "Charles S. Peirce at the Johns Hopkins," *Journal of Philosophy, Psychology, and Scientific Methods* 13 (1916): 715–22. Her remarks on Romanes come from "The Higher Education of Women," *The Century* 53 (1896): 315–16. A significant source for this chapter was Judy Green and Jeanne LaDuke, *Pioneering Women in American Mathematics: The Pre-1940 Ph.D.'s.* (Providence, RI: American Mathematical Society, 2009).

Chapter 9. Straddler

The epigraph is from Kelly Miller, *Out of the House of Bondage* (New York: Thomas Y. Crowell, 1914), 90–91. Miller describes his "first contact with Northern Culture" in his unpublished autobiography, available at Howard University's Moorland-Spingarn Research Center, Manuscript Division, Washington, DC. The later quote about illustrating fractions by cutting an apple is also from this source. Miller's comment on literacy laws comes from Kelly Miller, *Race Adjustment* (New York: Neale, 1909), 260. The "negro problem" is described in Frederick L. Hoffman, *Race Traits and Tendencies of the American Negro* (New York: Macmillan, 1896), v, 310–11. The Thomas Dixon article quoted is "Booker T. Washington and the Negro," *Saturday Evening Post* 175, no. 5 (August 19, 1905): 1–2. Miller's reputation as a "straddler" is described in August Meier, "The Racial and Educational Philosophy of Kelly Miller, 1895–1915," *Journal of Negro Education* 29 (Spring 1960), 121–27.

Chapter 10. Frontiersmen

The epigraph is from "Art of Compiling Statistics," specification forming part of Letters Patent No. 395,782, dated January 8, 1889. Simon Newcomb's comment about it being "pure nonsense" to talk of different sizes of infinity is from a letter he wrote to C. S. Peirce, March 9, 1892, quoted in David Lindsay Roberts, *American Mathematicians as Educators, 1893–1923: Historical Roots of the Math Wars* (Boston: Docent Press, 2012), 118. Grace Hopper's remark about punched cards is from Charlene W. Billings, *Grace Hopper: Navy Admiral and Computer Pioneer* (Hillside. NJ: Enslow, 1989), 61.

Chapter 11. Poetic Historian

The epigraph is from John Ewing's review of *The History of Modern Mathematics* in *Historia Mathematica* 19 (1992): 93. The reviews of Bell's book, *The Queen of the Sciences*, are by Paul H. Linehan, *American Mathematical Monthly* 39 (May 1932): 296–97, and David Eugene Smith, *Mathematics Teacher* 25 (April 1932): 238. Bell's remark about Leibniz is from E. T. Bell, *Men of Mathematics* (New York: Simon & Schuster, 1937), 120. He refers to Archimedes, Newton, and Gauss as being "in a class by themselves" on page 218 of that book. On mathematicians seeing themselves as "real persons" because of Bell's *Men of Mathematics*, see L. E. C., "Geometry and Mr. Newell," *Mathematics Teacher* 38 (December 1945): 347. E. T. Bell disparages "the merely competent" in *The Development of Mathematics* (New York: McGraw-Hill, 1940), 435. Freeman Dyson's recollection of the "club that you had to join" is from Donald J. Albers, "Interview with Freeman Dyson," *College Mathematics Journal* 25 (January 1994): 20. Bell's remarks on Jews can be found on pages 562–63 of the 1937 edition of *Men of Mathematics*. They persist in the 1961 Essandess paperback but have been sanitized in the 1965 Fireside paperback. All three editions have identical pagination. Steve Strogatz alerted me to this feature of the publishing history of *Men of Mathematics* (personal communication, January 20, 2014). "Witty and ironical" is the phrase of W. D. Reeve in his review of *Men of Mathematics* in *Mathematics Teacher* 31 (February 1938): 85. "Stark, frank humor" is remarked by G. Waldo Dunnington in his review of *Men of Mathematics* in *National Mathematics Magazine* 11 (May 1937): 406. "Cheap jokes" are lamented by George Sarton in his review of *Men of Mathematics* in *Isis* 28 (May 1938): 511. "Superlatively illuminating" is the assessment of Lao G. Simons in her review of *Men of Mathematics* in *American Mathematical Monthly* 45 (January 1938): 44. Bell's emphasis on myths is noted by G. A. Miller, "Mathematical Myths," *National Mathematics Magazine* 12 (May 1938): 388–89. Reuben Hersh's praise of Bell is found in "Let's Teach Philosophy of Mathematics!," *College Mathematics Journal* 21 (March 1990): 109. Ian Stewart's "cracking read" comment is quoted in Marion Cohen's review of *Letters to a Young Mathematician* in *American Mathematical Monthly* 115 (February 2008): 175. Gian-Carlo Rota hails the solution of Fermat's last theorem in *Indiscrete Thoughts* (Boston: Birkhauser, 1996), 141. A significant source for this chapter was Constance Reid, *The Search for E. T. Bell* (Washington, DC: Mathematical Association of America, 1993).

Chapter 12. Man of School Mathematics

The epigraph is from C. M. Austin, "Club History," May 1960, in the Records of the Men's Mathematics Club of Chicago, Box "Men's Mathematics Club Early History Pre 1930–1950s," Folder "Early History (prior to 1930)," Northeastern Illinois University Archives. Capitalization in original. The assertion that "the less we have to do with the women the better" is from a letter from O. E. Overn to F. W. Runge, May 2, 1932, in the Records of the Men's Mathematics Club. Ella Flagg Young's complaint that mathematics was "too mechanical" comes from John T. McManis, *Ella Flagg Young and a Half-Century of the Chicago Public Schools* (Chicago: A. C. McClurg, 1916), 75. The quotes from the Mathematics Club committee mentioned are from Alfred Davis, "The Status of Mathematics in the Secondary Schools," *School Science and Mathematics* 18 (January 1918): 25–35, and Alfred Davis, "Valid Aims and Purposes for the Study of Mathematics in Secondary Schools," *School Science and Mathematics* 18 (February–April 1918): 112–23, 208–20, 313–24. C. M. Austin's comments on testing are from his "An Experiment in Testing and Classifying Pupils in Beginning Algebra," *Mathematics Teacher* 17 (1924): 46–56. Austin's remarks in Miami in 1964 were recalled by James D. Gates, oral history interview with David L. Roberts, March 13, 2004, National Council of Teachers of Mathematics Oral History Project Records, 1992–93, 2002–4, Archives of American Mathematics, Center for American History, University of Texas at Austin. A significant source for this chapter was Glenn Hewitt, "History of the Men's Mathematics Club of Chicago," in *A Half Century of Mathematics Progress*, edited by David Rappaport (Chicago: privately published, 1965).

Chapter 13. Organization Man

The epigraph is from Paul A. Samuelson, "Gibbs in Economics," in *Proceedings of the Gibbs Symposium*, edited by D. G. Caldi and G. D. Mostow (American Mathematical Society, 1990), 259. Wilson's suggestion to give a difficult job to "an able young person" is from an interview of Edwin Wilson by R. Bruce Lindsay and W. J. King on June 3, 1964, Niels Bohr Library and Archives, American Institute of Physics, College Park, Maryland. Wilson's complaints about "schizophrenic" physics is found in Edwin B. Wilson, "What Is Statistics?," *Science*, n.s., 65 (June 17, 1927): 586. Wilson's conservative economic comments come from Edwin Bidwell Wilson's review of Vilfredo Pareto, *Manuel d'Economie politique*, *Bulletin of the American Mathematical Society* 18 (1912): 473–74. The Samuelson quote on "slippery slopes" is from his obituary in *The Economist* (December 17, 2009). Wilson's comments on "the mild little cigarette" come from Edwin B. Wilson, "The Cigarette–Lung Cancer Enigma," talk presented before small group in Boston, January 4, 1960, Industry Documents Library website, http://industrydocuments.library.ucsf.edu /tobacco/docs/fyby0141. "Whipping boy" and "substantial good" are from E. B. Wilson to C. C. Little, February 5, 1961, Industry Documents Library website, http://industrydocuments.library.ucsf.edu/tobacco/docs/hxhw0216. A significant source for this chapter was Jerome Hunsaker and Saunders Mac Lane, "Edwin Bidwell Wilson," in *National Academy of Sciences Biographical Memoirs* (Washington, DC: National Academy of Sciences, 1973).

Chapter 14. Versed in Math

The epigraph is from Lillian R. Lieber, *Infinity: Beyond the Beyond the Beyond* (Philadelphia, PA: Paul Dry Books, 2007), 257. Lillian Lieber's writing is described as "blank verse" in Mary Thompson, "Mathematics Aids Couple over Matrimonial Rocks," *Pittsburgh Press*, July 17, 1938, and as "free verse" in H. W. Brinkmann, "Recent Publications," *American Mathematical Monthly* 41 (August–September 1934): 442–43. Daughters were described as "empty nutshells" in Miriam Shomer Zunser, *Yesterday: A Memoir of a Russian Jewish Family* (New York: Harper & Row, 1978), 65. Page 141 of this book describes the "lopped off" existence of Lillian's parents in New York. The Jewish threat to "homes of refinement" is mentioned in Jerome Karabel, *The Chosen: The Hidden History of Admission and Exclusion at Harvard*,

Yale, and Princeton (Boston: Houghton Mifflin, 2005), 87. Edward Kasner introduced "googol" in his "New Names in Mathematics," *Scripta Mathematica* 5 (1937): 13. John Gunther quotes his son on the Liebers in *Death Be Not Proud* (New York: Harper & Row, 1949), 135.

Chapter 15. Machine Whisperer

The epigraph comes from an interview with Grace Murray Hopper by Uta C. Merzbach (July 1968) in the Computer Oral History Collection, 1969–73, 1977, Archives Center, National Museum of American History, Smithsonian Institution, Washington, DC. This interview is also the source for Hopper's comment about not needing to figure out what to cook for dinner when she joined the navy. The Jaron Lanier quote is from his introduction to Ellen Ullman, *Close to the Machine: Technophilia and Its Discontents* (New York: Picador, 1997). Two significant sources for this chapter were Charlene W. Billings, *Grace Hopper: Navy Admiral and Computer Pioneer* (Hillside, NJ: Enslow, 1989), and Kurt W. Beyer, *Grace Hopper and the Invention of the Information Age* (Washington, DC: Smithsonian Institution, 2009).

Chapter 16. Survivor

The epigraph is from Izaak Wirszup, oral history interview by David L. Roberts, November 28, 2000, R. L. Moore Legacy Collection, 1890–1900, 1920–2003, Archives of American Mathematics, Center for American History, University of Texas at Austin. This is also the source for Wirszup's assessment of Marcinkiewicz as a "genius type." Wirszup talked of "broadening and vivifying the student's significant mathematical experience" in "Some Remarks on Enrichment," *Mathematics Teacher* 49 (November 1956): 519. This is the published version of his Milwaukee talk of 1956. The appreciation of Izaak and Pera's salon comes from a personal communication from Alphonse Buccino, April 23, 2000.

Chapter 17. Carrying Old Virginny Forward

The title of this chapter was inspired by reading L. Douglas Wilder, *Son of Virginia: A Life in America's Political Arena* (Guilford, CT: Rowman & Littlefield, 2015). Wilder's attendance at Virginia Union partly intersected with that of Edgar Edwards. As governor of Virginia in the 1980s, Wilder criticized the racist lyrics of "Carry Me Back to Old Virginny" and campaigned to dethrone it as the official state song. The epigraph comes from James Jackson Kilpatrick, *The Southern Case for School Segregation* (New York: Crowell–Collier Press, 1962), 89. All direct quotes from Edwards come from Edgar L. Edwards Jr., oral history interview with David L. Roberts, October 17, 2002, National Council of Teachers of Mathematics Oral History Project Records, 1992–93, 2002–4, Archives of American Mathematics, Center for American History, University of Texas at Austin. The mission statement for the Benjamin Banneker Association comes from its website, accessed August 10, 2018, http://bbamath.org/.

Chapter 18. Americano

The epigraph is from J. B. Diaz, "A Comparison of Two Uniqueness Theorems for the Ordinary Differential Equation $y' = f(x,y)$," *Inequalities* 3 (1972): 65. The "Americano" anecdote is from a personal communication from Diaz's daughter, Joan Diaz, August 2, 2018. "Do I look like Albert Einstein?" comes from a June 18, 2018, communication from Joan. She was also the source for "rolling in the aisles," date uncertain. The remark on R. L. Moore's views on what constituted "really American" is from an interview with Martin Ettlinger, a classmate of Joe Diaz at the University of Texas (UT), and the son of Hyman Ettlinger, Diaz's professor. The transcript of this interview, still not fully processed by the Center for American History at UT, was kindly provided by Albert Lewis of the Educational Advancement Foundation. The Lipman Bers quip about "heavily accented English" can be found in Irwin Kra and Hyman Bass, "Lipman Bers, 1914–1993," in *Biographical Memoirs*, vol. 80 (Washington, DC: National Academy of Sciences, 2001), 14. Diaz's remark on "the gradual process of unification and simplification" is from Joaquin B. Diaz, "How Mathematics Progresses," *Historia Mathematica* 2 (1975): 601. The Paul Cohen quote comes from Sylvia Nasar, *A Beautiful Mind: The Life of Mathematical Genius and Nobel Laureate John Nash* (New York: Touchstone, 1998), 237. Much of the biographical information about Diaz comes from primary documents and personal communications collected during 1992–93 by Florence Fasanelli, working on behalf of the Mathematical Association of America.

Chapter 19. Math Warrior

The epigraph is from Frank B. Allen, "The New Math—An Opportunity Lost," *Mathematics Teacher* 77 (November 1984): 590. Allen discusses tracking into "minimum," "regular," and "superior," in "Building a Mathematics Program—An Adventure in Co-Operative Planning," *Mathematics Teacher* 49 (April 1956): 229. All other direct quotes are from Frank B. Allen, oral history interview by David L. Roberts, August 12, 1999, R. L. Moore Legacy Collection, 1890–1900, 1920–2003, Archives of American Mathematics, Center for American History, University of Texas at Austin. Max Beberman's warnings about "charlatans" and "scandal" are from Harry Schwartz, "Peril to Doing Sums Seen in 'New Math,'" *New York Times*, December 31, 1964.

Chapter 20. Suspicious Minds

The epigraph comes from William Barrett, *Irrational Man* (Garden City, NY: Doubleday Anchor, 1962), 84. Lynne M. Butler's assessment of Russell Crowe's performance is from her article "A Beautiful Mind," *Notices of the American Mathematical Society* 49 (April 2002): 455–57. The Sylvia Nasar quotes are all from her book *A Beautiful Mind: The Life of Mathematical Genius and Nobel Laureate John Nash* (New York: Touchstone, 1998). Serge Lang is quoted in Michael Fitzgerald and Ioan

James, *The Mind of the Mathematician* (Baltimore: Johns Hopkins University Press, 2007), 7. The *Fortune* article that mentions Nash is George A. W. Boehm, "The New Uses of the Abstract," *Fortune* 58 (July 1958): 124–27, 152, 156, 158, 160.

Conclusion

The epigraph comes from Edwin B. Wilson, "Statistics Then and Now," *American Statistician* 19 (April 1965): 38. This is the text of a talk given by Wilson one month before his death, in December 1964. Thurston's remark is from W. P. Thurston, "Mathematical Education," *Notices of the American Mathematical Society* 37 (September 1990): 848. The prediction of shirt-pocket computers was made by John B. Conway, "A Wealth of Potential but an Uncertain Future: Today's Mathematics Departments," *Notices of the American Mathematical Society* 44 (April 1997): 439. The accidental predominance of partial differential equations was remarked in Stephen Wolfram, *A New Kind of Science* (Champaign, IL: Wolfram Media, 2002), 162. His further remarks about the future of mathematics are from the same source, page 821.

Index

Page numbers in *italics* refer to illustrations.